Asbestos

and its implications for members and their clients

Published by RICS Business Services Limited
a wholly owned subsidiary of
The Royal Institution of Chartered Surveyors
under the RICS Books imprint
Surveyor Court
Westwood Business Park
Coventry CV4 8JE
UK

No responsibility for loss occasioned to any person acting or refraining from
action as a result of the material included in this publication can be accepted
by the author or publisher.

ISBN 1 84219 063 6

HSE material reproduced with kind permission of the HSE

Reprinted 2003

Typeset in Great Britain by Columns Design Limited, Reading
Printed in Great Britain by Bell & Bain Ltd, Glasgow

Contents

Preface

Asbestos is the most serious occupational health issue, in terms of fatalities, that the UK has ever faced. Asbestos was used so widely in construction products until comparatively recently that almost anyone carrying out building or maintenance work is potentially at risk from exposure to asbestos fibres. This is why the government recently brought in new legislation, the *Control of Asbestos at Work Regulations* 2002, to require those with responsibility for any maintenance activities to take effective action to manage the risk from asbestos in their buildings.

This 'duty to manage' is not about removing all asbestos from buildings. Rather, it is about finding where the asbestos is present, assessing the risk and, depending on the condition of the asbestos and whether it is likely to be disturbed, taking action to manage that risk – both in the short and long term.

The Health and Safety Executive (HSE) estimates that up to 500,000 commercial, industrial and public buildings have asbestos materials in them. All those responsible for buildings need to be made aware of the risks arising from asbestos and of how to comply with the 'duty to manage' in a way that is proportionate to the risks. This will be a considerable challenge, and one that cannot be met without the assistance of all interested parties. I believe that the surveying profession has a key role to play in helping clients to make the right decisions – in what for many of them will be a difficult area.

This guidance note demonstrates how seriously RICS takes the threat of asbestos. The HSE is pleased to be associated with it. Indeed, we look forward to co-operating ever more closely with RICS on asbestos and other health and safety issues in the future.

John Thompson
Head of Chemicals and Flammables Policy Division
Health and Safety Executive

RICS guidance notes

This is an RICS guidance note. It provides advice to members of RICS on aspects of the profession. Where procedures are recommended for specific professional tasks, these are intended to embody 'best practice', that is, procedures which in the opinion of RICS meet a high standard of professional competence.

Members are not required to follow the advice and recommendations contained in the guidance note. They should, however, note the following points.

When an allegation of professional negligence is made against a surveyor, the court is likely to take account of the contents of any relevant guidance notes published by RICS in deciding whether or not the surveyor has acted with reasonable competence.

In the opinion of RICS, a member conforming to the practices recommended in this guidance note should have at least a partial defence to an allegation of negligence by virtue of having followed those practices. However, members have the responsibility of deciding when it is appropriate to follow the guidance. If it is followed in an inappropriate case, the member will not be exonerated merely because the recommendations were found in an RICS guidance note.

On the other hand, it does not follow that a member will be adjudged negligent if he or she has not followed the practices recommended in this guidance note. It is for each individual chartered surveyor to decide on the appropriate procedure to follow in any professional task. However, where members depart from the good practice recommended in this guidance note, they should do so only for good reason. In the event of litigation, the court may require them to explain why they decided not to adopt the recommended practice.

In addition, guidance notes are relevant to professional competence in that each surveyor should be up to date and should have informed him or herself of guidance notes within a reasonable time of their promulgation.

Other sources of guidance

In addition to RICS guidance notes, this document contains references to and extracts from other RICS publications, including the RICS *Appraisal and Valuation Standards* (the *Red Book*), as well as various documents published by the Health and Safety Executive (HSE).

In some cases, the information contained in these publications is only advisory, while in others it has special professional or legal status. It is therefore necessary to explain the hierarchy of these publications.

RICS practice statements

The RICS *Appraisal and Valuation Standards* (the *Red Book*, 5th edition) (see section 5.8) contains practice statements, guidance notes and appendices.

The practice statements have mandatory status and should be applied to all valuations and assessments of worth provided by members, in all states and for all purposes to which the standards apply. They do not apply to valuations for certain defined purposes, which are as follows: advice given during the course of litigation, arbitrations and similar disputes, or during negotiations; internal valuations; certain agency or brokerage work; development schemes; and with regard to antiques, fine art and chattels.

Where a member considers that there are special circumstances that make it inappropriate or impractical for a valuation to be made wholly in accordance with the relevant practice statements, those circumstances must be confirmed and agreed with the client as specific departures before reporting. The adoption of a special assumption is not regarded as a departure. A clear statement in writing of any departures, together with details of, and the reasons for these, and the client's agreement, must be given in the report.

Hierarchy of HSE publications

This document makes reference to regulations, Approved Codes of Practice (ACOP) and other publications produced by the HSE. An explanation of their hierarchy and legal significance is set out below.

Regulations

These set out the legal requirements that will be considered by the courts as a first point of reference to establish compliance with statute.

Approved Codes of Practice (ACOP)

These provide advice on the preferred means of compliance with specific regulations.

They have a special legal status, such that in the event of a prosecution for a breach of a regulation, where it is proved that the relevant provisions of the ACOP were not followed, the court will find the accused guilty unless it can be shown that the law was complied with in some other way.

Guidance

HSE guidance gives more general advice on ways of complying with the law, but does not have the special legal status of the ACOP.

Foreword

Asbestos can be found in all but the newest of buildings.

It has an impact on the majority of RICS members in one way or another, with significance for almost all of the professional services that we provide. In addition, it presents a potentially serious risk to our own health and that of our colleagues, employees and clients.

Whether our involvement with asbestos is specific – actually looking for, or advising on asbestos – or only incidental; whether we provide the service ourselves, or employ or advise on the use of specialist contractors or consultants; we need to understand the key issues associated with this insidious material, the regulations that apply and the guidance that is available.

To quote an extract from the European Parliament directive on the protection of workers from the risks of exposure to asbestos:

> 'A well informed worker will be more cautious in carrying out its activities'.

This is of particular relevance to the surveyor, where caution is required in understanding the real risks, commercial and financial, as well as for health, relating to asbestos, and in ensuring that advice or action is appropriate and commensurate with these.

In the case of asbestos, exaggeration can be as damaging as underestimation, and a balanced, unemotional and professional approach is essential.

Part 1
Introduction

1.1   The context

Asbestos, in its nature and effects, has been compared to witchcraft in the Middle Ages. It conjures up irrational fears and often provokes inappropriate and excessive reactions. The whole subject is seemingly shrouded in the mystique of the 'black arts', the secrets of which are confined to a few specialized 'practitioners'.

Asbestos is commonly found in all but the most recent of premises of almost any kind. It can present a serious, sometimes fatal, risk to the health of anyone, including surveyors, who disturbs it, intentionally or inadvertently.

Asbestos is anathema to the insurance industry. The level of claims relating to asbestos (predominantly from the US market) has caused some insurers to stop trading, or to move out of this specialized market altogether, and has heightened the caution of those remaining.

Asbestos is at the top of many of the lists of 'deleterious materials' prepared by property owners and investors. Its very presence, irrespective of the level of risk to health it presents, can have an adverse effect on the value of buildings, plant and machinery – or their liquidity.

Asbestos is also probably one of the most highly regulated topics in the UK. There is a plethora of regulations, codes of practice and official guidance dealing with its use, disturbance, treatment and removal. The legislation has developed to address specific issues or concerns and often it is necessary to consider a number of documents concurrently, in order to understand the full impact of a particular situation.

To the novice, the whole subject may appear complicated and daunting, discouraging further interest or investigation and prompting the delegation of asbestos problems to the few 'professional experts' or to asbestos contractors.

This guidance note is intended to give readers a balanced and pragmatic appreciation of the various issues surrounding asbestos, with particular regard to its use in buildings and structures.

1.2   Purpose of this document

Asbestos can affect surveyors and their clients in a variety of ways – as employers, employees, advisors, owners, investors in or occupiers of buildings.

This document is pertinent to clients commissioning surveys or the repair, treatment or removal of asbestos; surveyors carrying out or organizing inspections of buildings containing the material; and persons instructing contractors or briefing other specialists. It is intended to provide such persons with a general understanding of the issues, to enable them to protect themselves and others who may be affected by their activities, both personally and commercially.

It is primarily intended for the majority of surveyors, who will not have specialist knowledge and expertise in respect of asbestos, but who provide a diverse range of property services, including management and valuation, agency, inspection and the provision of professional advice on a wide variety of building types and tenures.

It is essentially a starting point, designed to provide a foundation in the subject. Where the service provided requires a more detailed knowledge of asbestos, it aims to direct the surveyor to the abundance of legislation and official guidance.

Its aims are as follows.

a) Generally, to:

- give a basic awareness of the types and uses of asbestos commonly found in buildings and the possible locations of these;

- explain the risks to health and the possible implications of the presence, removal or treatment of asbestos, and how to assess these;

- summarize the key regulations and legislation;

- explain the role of the various specialist parties and organizations involved in the asbestos industry, including asbestos inspectors, analysts, licensed asbestos removal contractors and the regulatory authorities.

b) More specifically, for the surveyor, to:

- demonstrate how asbestos may impact on the various professional surveying services;

- clarify professional responsibilities (where not defined by specific instructions), having regard to current law and the limitations of professional indemnity insurance;

- assist with the consideration of specialists' reports, asbestos registers and management plans;

- recommend standard reporting phrases for use in particular circumstances;

- consider the factors that need to be taken into account when contemplating carrying out, or commissioning an inspection for asbestos, including sampling requirements and the preparation of a plan to manage asbestos.

1.3   Other documents referred to

The guidance contained within this document is based on the current editions of other guidance notes and publications produced by RICS and the Health and Safety Executive (HSE), as well as relevant case law. It considers existing regulations and official codes of practice.

At the time of writing, asbestos issues are at the forefront of the concerns of both the British government and the European Commission. It is likely that there will be further legislative changes that may necessitate the amendment of this or other RICS guidance.

It is therefore hoped that this format will enable the reader to make appropriate alterations to the text, to reflect any subsequent significant changes. In this way, the guidance may be updated as necessary, until such time as revised documents are published.

As in all circumstances where the law is concerned, the final arbiter will be the courts. However, in the absence of relevant case law, surveyors must rely on informed opinion. Expert legal opinion has been sought in the matters of liability and implications for surveyors – in particular, in respect of their limitations and exclusions.

A bibliography is contained in Appendix 9.

1.4 Structure of the document

This document has been laid out in a logical format, to enable it to be read in the usual way, from cover to cover. In addition, a number of devices have been incorporated to facilitate easy reference for retrieval of information on specific topics, if required.

As the guidance is intended for readers with a wide variation in their experience, appreciation of and interest in asbestos, the general text essentially summarizes the key issues. It refers to more detailed information, charts and tables contained in the appendices.

A section has also been included for each of the various professional services provided by RICS members, to demonstrate how asbestos may impact on their particular specialisms, giving examples and case studies where appropriate.

1.5 *Control of Asbestos at Work Regulations*

The key legislation in respect of works associated with or involving 'asbestos-containing materials' (ACMs) are the *Control of Asbestos at Work Regulations (CAWR)*.

At the time that this guidance was drafted, following a long period of consultation, the Regulations, which originated in 1987, together with Amendment Regulations in 1987 and 1992, had only recently been consolidated into one set of Regulations – *The Control of Asbestos at Work Regulations* 2002.

An important amendment that was added, with particular and significant implications for members of RICS and their clients, was Regulation 4, the 'Duty to manage asbestos in non-domestic premises'.

The other significant change was Regulation 20, 'Standards for analysis': this states that the analysis of samples of materials to determine whether they contain asbestos may only be undertaken by a person accredited as complying with ISO 17025 by the United Kingdom Accreditation Service (UKAS).

With the exception of Regulations 4 and 20, *CAWR* 2002 came into force on 21 November 2002. Because of the likely impact on both the asbestos and property industries, it was deemed necessary for these two Regulations to have a transitional period, following their publication, before they came into force. Consequently, Regulation 4 will come into force on 21 May 2004 and Regulation 20 on 21 November 2004.

While before these dates there will be no legal requirement to comply with these two regulations, the HSE and the government are anxious that the asbestos and property industries should start to adopt the principles and procedures of the Regulations as early as possible. The dates for the Regulations to come into force should be considered as 'drop dead' dates, by which time every 'dutyholder' will be expected to be up to speed, and to have embraced and instigated the changes in their working methods needed to comply with the Regulations.

In some cases, due to the number or size of buildings, or the lack of available resources, including the availability of expert advice, there may be justification for failure fully to implement the plans required under Regulation 4 by 21 May 2004. As a minimum, however, the HSE will expect the building(s) to have been visited and reviewed; priorities to have been established based on the assessed risk (the likelihood of incorporating ACMs and/or their condition); a realistic programme to have been prepared; and a budget to have been allocated. (See Appendix 4 'Regulations' for details of Regulation 4.)

RICS members are encouraged immediately to adopt these two Regulations as good working practice, and to encourage a similar response from those who employ them and to whom they provide advice.

It should be appreciated that although Regulation 4 sets out specific procedures to be followed and documents to be produced, there are also existing, and more general, health and safety regulations with the same objectives. In the event of breach of these general regulations, criminal prosecution may follow.

Similarly, Regulation 20 simply makes mandatory the good practice that is currently followed by the majority of professional asbestos analysts. There are also numerous general regulations that already require employers to ensure that anyone appointed is competent and properly resourced. The new Regulation therefore merely adds definition to the necessary standards.

In view of RICS' commitment to support the HSE in its objectives, this guidance note has been written as if all of the *CAWR*, including Regulations

4 and 20, were already in force. However, there are a number of instances where it is appropriate that the current legal situation during the transitory period of Regulations 4 and 20 is considered, and in such cases this is made clear in the text. Overall, unless specifically noted to the contrary, any reference to *CAWR* is in respect of the full 2002 Regulations, as will be in force from 21 November 2004.

1.6 Asbestos survey or inspection…surveyor or inspector?

Legislation and guidance on asbestos, both specific and general, has been published by various regulatory and professional bodies, but with little commonality, especially in the terminology employed.

In such documents asbestos is variously referred to as a deleterious material, a hazardous material, a contaminant and a source of contamination. All of these terms can legitimately apply to asbestos, but confusion can arise as they often have a very specific meaning, which can differ subtly depending on the document.

See Appendix 1 'Definitions and abbreviations used' for definitions of these and other words as used in this document.

Other associated words that are frequently used in connection with asbestos are 'surveyor', 'inspector', 'survey' and 'inspection'. In order to avoid similar confusion, in this guidance note the following meanings are used.

Surveyor

A surveyor is a member of RICS, or a similar professional organization, who carries out one or a combination of professional services that involve the 'survey' of buildings (e.g. a building surveyor, a general practice surveyor or a valuer).

The 'survey' of buildings can be for a variety of purposes, but within it, asbestos will be just one of a number of matters to be taken into consideration, and not the prime objective of the service provided. The 'surveyor's' involvement with asbestos is thus only incidental, as part of the general service provided.

Asbestos inspector

An asbestos inspector is a person or organization who conducts an 'asbestos inspection' of buildings, structures, plant or land, with the specific and single objective of identifying and reporting on ACMs.

The information could be required for a number of purposes, such as the compilation of an 'asbestos management plan', or an 'asbestos register' or to determine the existence or otherwise of ACMs prior to demolition or alterations of buildings, plant or equipment that might otherwise disturb ACMs.

At present, there is no mandatory qualification for 'asbestos inspectors' and anyone, including 'surveyors', can provide the service, provided that they are competent and have the necessary specialist knowledge and experience appropriate for the specific project.

The term 'asbestos inspector' is not widely used at present, but future personal certification measures may adopt this designation. (See Appendix 7 for further details.)

The level of expertise on asbestos-related matters that can be expected of an 'asbestos inspector' (commissioned for a project that is only and specifically concerned with asbestos) is far greater than that reasonably required of a 'surveyor', who is obliged to comment on asbestos incidentally in the generality of numerous other topics.

Where a member of RICS offers both services, it is important that in so far as asbestos is concerned, the client is made aware of the scope of the specific service that will be provided, either as a 'surveyor' or an 'asbestos inspector', and is clear as to the difference between a 'building survey' and an 'asbestos inspection'.

Asbestos survey

Numerous HSE publications, including in particular the guidance document MDHS 100 *Surveying, sampling and assessment of asbestos-containing material,* use the term 'survey' in relation to investigations solely concerning asbestos. However, this term is historically and by definition linked to RICS members, and has specific meaning to them and their clients in the context of the various different services they provide.

Therefore, in order to avoid confusion, this guidance note does not use the term 'asbestos survey', except where referring to specific HSE publications in which it is mentioned, and hereafter uses the more appropriate term 'asbestos inspection'. This distinguishes this very specific service carried out by 'asbestos inspectors' from that of the range of services provided by 'surveyors'.

It is strongly suggested that the terms 'asbestos inspection' and 'asbestos inspector' are used by RICS members whenever appropriate, to encourage the common usage of this new terminology.

This has particular relevance in respect of professional indemnity insurance (see section 7, 'Insurance').

Asbestos inspection

An asbestos inspection is an inspection of buildings, structures, plant and land, where the sole objective is to determine or assume the location, type and condition of materials containing asbestos. (See Section 8 'Asbestos inspection' for further details.)

1.7   Abbreviations frequently used

For brevity, the following abbreviations have been used for phrases, names and important regulations that occur frequently in the text.

For a full list of definitions and abbreviations used, see Appendix 1.

ACM   Asbestos-containing material

CAWR   *Control of Asbestos at Work Regulations* (2002, unless otherwise stated)

HSE   Health and Safety Executive

Part 2
Risk to RICS members

2.1 General

The general health issues associated with asbestos for the public at large are summarized in Appendix 2 'Health issues and risks arising from asbestos'. This section addresses the risks from the perspective of the surveyor.

The risks arising from asbestos are broadly of two types: asbestos-related diseases and commercial/economic loss.

(Refer to the list in Appendix 1 for the definition of hazard and risk, to gain an understanding of the subtle difference between the two.)

2.2 Asbestos-related diseases

Two factors must be taken into account here:

● the health of surveyors; and

● the health of others (occupants, visitors, neighbours, contractors and the general public).

The term 'asbestos-related diseases' is used here to refer to illness or death arising from the inhalation of asbestos fibres. Inhalation or ingestion may occur during the inspection, maintenance, alteration or demolition of components, plant, equipment or buildings containing asbestos-containing materials (ACMs). Such activities, without adequate control, can release significant quantities of respirable airborne asbestos fibres.

2.3 Commercial/economic loss

The following are included under this heading:

● emergency or unplanned stoppage of production and/or cessation of services;

● evacuation of a building or parts thereof, including the costs of the provision of temporary alternative accommodation and facilities;

● loss of immediate income, due to closure or boycott by customers (for example, in the case of cinemas, theatres or shops);

● strikes or walkouts by employees or occupants;

● adverse publicity (for example, for blue-chip companies or schools);

● reduction in value or rental income;

- loss of liquidity of asset (difficulty in selling, leasing or licensing the premises or inability to do so);

- costs of remedial works (removal or treatment and decontamination);

- financial responsibility for injured employees or other parties;

- criminal prosecution (leading to substantial fines and even imprisonment); and

- civil damages for negligence.

In considering loss, it must not be forgotten that similar or greater commercial economic risk can arise from taking an overly cautious and unrealistic stance as from ignoring the problem. Such an approach could result in unnecessary fear and concern, and in needless remediation or removal works. There is the possibility of criminal prosecution if persons are exposed to the risks of asbestos by unnecessary works.

The surveyor must always take a balanced and professional view dependent upon the individual circumstances and must not 'sit on the fence' or 'take the easy option'. He or she must tread a fine line between caution and pragmatism, relying on the published guidance that is available.

Weight must be given to information produced by relevant professional organizations and institutions, such as RICS, official authoritative government departments and the Health and Safety Executive (HSE), rather than private organizations that may have a commercial interest and bias. As an example, there is currently a debate as to the level of risk arising from white asbestos. For the avoidance of doubt, see Appendix 6 'Legal issues arising and cases' for the HSE's view on this.

Part 3

Surveyors' responsibilities

3.1   General duty

Members of RICS provide a wide range of property services, each with different motivations and objectives, and each requiring very different levels of skill and knowledge. Quite correctly, these differences are reflected in members' attitudes towards asbestos, with regard both to the level of risk that it presents to them personally and the extent to which they are involved with it in their everyday activities.

Irrespective of personal views and the level of involvement, all RICS members, with all of the moral and ethical responsibilities that membership entails, have specific obligations set out in the Institution's Royal Charter.

> ' ... the objects of the Institution shall be to secure the advancement and facilitate the acquisition of that knowledge which constitutes the profession of a surveyor, namely, the arts, sciences and practices of ... securing the optimal use of land and its associated resources to meet social and economic needs ... and to maintain and promote the usefulness of the profession for the public advantage'.

The government has repeatedly expressed its concern regarding the protection of workers and the general public from risks to health arising from asbestos in buildings, and RICS has on many occasions confirmed its willingness to assist in this in every reasonably practicable way.

As a minimum, every RICS member, whether acting in the capacity of an inspector, owner, manager, occupier of or adviser on buildings and land, must be aware of the implications of asbestos, the statutory obligations imposed on various parties with relation to asbestos, and the regulatory requirements, in order to be able to provide the necessary professional and impartial advice to enable competent expert assistance to be sought and given.

To many people, including some surveyors, asbestos is a material to be avoided wherever possible. Because of the perceived risks, there may be a tendency for surveyors to try to evade or opt out of any involvement with it whatsoever. Because of professional and statutory obligations, however (as explained in detail later in the text), this is not possible. Surveyors must therefore face up to their responsibilities.

Many surveyors will appreciate the commercial advantages of taking the lead in providing advice or services in connection with a matter that has such serious financial implications for the property world. Others should be aware of the potential commercial disadvantages of not doing so.

In view of the size of RICS membership, and the significant and prestigious positions held by many members in the property world, it is unlikely that the national and EC objective of managing the risk of asbestos within the workplace can be achieved within the desired timescale, or even at all, without the support, commitment and influence of RICS members.

3.2 Statutory responsibility

Irrespective of financial and commercial considerations, the surveyor's prime concern must be the risk to health and of bodily injury.

Notwithstanding contractual obligations and conditions of engagement, section 3 of the *Health and Safety at Work etc. Act* 1974 requires everyone at work (including surveyors) to do that which is reasonably within their control to prevent the injury of others. (This embraces 'passive' as well as active responses – i.e., including the failure to warn or act.)

Accordingly, if during the course of an inspection, a surveyor identifies or reasonably suspects the presence of materials that may contain asbestos, and where the risk to health is considered to be 'immediate and significant', he or she should report this, irrespective of the conditions of engagement.

The criteria must be left to the individual to decide in the particular circumstances. An example of 'immediate and significant' risk would be the existence of substantial debris from materials reasonably suspected of or known to contain asbestos, with the potential for releasing airborne fibres if disturbed, in an area in which unprotected persons are present or are likely to enter in the near future.

Ideally, the person in control of the premises should also be informed and given advice on emergency or other action to be taken, which as a minimum should include seeking immediate advice from a specialist.

Where this could breach client confidentiality, it is suggested that this duty may be discharged by informing the client, clearly pointing out the need for the occupants to be informed as soon as possible and recording the notification. The duty to inform the appropriate parties will then pass to the client.

The only exception to the foregoing is where asbestos-containing materials (ACMs) are specifically excluded from the surveyor's inspection and the surveyor is aware that others have either carried out or will imminently be conducting an 'asbestos inspection', and this is noted in the terms of engagement.

3.3 Contractual responsibility

The scope of a surveyor's contractual responsibility will depend on the client's specific requirements. These must be made clear in the instructions received and confirmed prior to providing the service. (See sub-section 8.4 'Briefing for an inspection' for further advice.)

The requirements should be as detailed as possible, so that the expectation of both parties is clear. Where this is not so, case law suggests that the courts will take into account the status of the instructing parties, differentiating between the naive domestic client and the informed commercial manager or owner of a portfolio of properties, and will tend to give greater sway to the 'man in the street'.

RICS has published various guidance notes in this area, including *Building Surveys and Inspections of Commercial and Industrial Property*, *Building Surveys of Residential Property*, *Contamination and Its Implications for Chartered Surveyors* and the *Appraisal and Valuation Standards* (the *Red Book*).

The guidance notes set out advice on best practice (see 'RICS guidance notes' for further details of the legal significance of these), while the *Red Book* establishes mandatory requirements. All of these publications include specific and indirect references to asbestos (see section 5 'Services offered by surveyors' for relevant extracts from these publications). There can therefore be no doubt that, just like other common defects and concerns in buildings, surveyors are expected to have sufficient knowledge of asbestos and its possible implications, according to the type of service that they are providing.

Similarly, just as surveyors are reasonably expected to be aware of the peculiar and specific defects associated with a particular form of construction, for example, timber-framed or system-built buildings, they are expected to be conversant with the likely uses of asbestos commonly associated with the age and type of properties that they inspect. It is only the extent of the knowledge reasonably required that is in question.

Unless specifically stated or inferred by the client's brief, or associated correspondence, a surveyor will be judged on whether he or she has acted with the skill and care that could reasonably be expected of a reasonably competent surveyor in similar circumstances offering similar services. If a surveyor is putting him or herself forward, either directly or indirectly, as an 'asbestos inspector', with specific and special skills in this field, a greater degree of knowledge and expertise on asbestos may reasonably be expected.

In considering this, it must be remembered that the standards of the reasonably competent surveyor are not static, but develop as technical knowledge is enhanced and standards of technical competence are improved.

In determining whether a surveyor has acted with reasonable skill and care, the court will take notice of the circumstances surrounding the original survey, and the knowledge, general guidance and information available to the surveyor at the time the service was provided. He or she will then be judged by the comparable standards of a reasonably competent surveyor at that time.

It should be noted that with regard to asbestos, the benchmark will change with the publication of this guidance note.

Part 4

Responsibility of others (employer, landlord, tenant, occupant)

4.1 General

There are a large number of regulations, both specific and general, that apply to various parties according to the role(s) they perform and the particular circumstances of each case.

Many of the requirements are duplicated: for example, there is a requirement to carry out risk assessments generally under the *Management of Health and Safety at Work Regulations* 1999, and specifically under the *Control of Asbestos at Work Regulations* (*CAWR*). In such cases, compliance with the latter will satisfy the requirements of the former, in so far as this relates to the risk from asbestos-containing materials (ACMs).

The following section summarizes the aggregate responsibilities imposed by these numerous regulations, as they apply to employers, landlords, tenants, managers and occupants of buildings. Reference should be made to Appendix 4 'Regulations' for specific details of particular legislation.

Many of the issues arising will have implications and will thus be relevant for several parties concurrently. For brevity and to avoid repetition, in the following section not all of the relevant information is repeated for each role. For example, reference to delegating duties to others is referred to in the sub-section headed 'Landlord', but could equally apply to the sub-section headed 'Tenant'. It is therefore necessary to read and understand the responsibilities and issues arising in connection with each role to appreciate the full picture, as there will be occasions when a party will be performing a number of these roles at the same time.

4.2 Employer

In health and safety regulations, 'employer' is the term used to describe anyone who employs another person to perform a service, and to self-employed persons as well. The term applies to the 'employer' of workers who carry out operations that could disturb asbestos, as well as the 'employer' of those occupants of the premises who may be put at risk by these activities in the workplace.

Every employer has general responsibilities for the health and safety of employees and of other persons, and must:

● conduct their works in such a way that employees and others are not exposed to health and safety risks;

- be aware of hazards to health and safety arising from the work activity or workplace, assess the risks arising and take appropriate measures, as reasonably practicable, to remove, reduce or control these; and

- provide information to employees and other persons about the work or workplace that might affect their health and safety.

In addition, an employer may also be a 'landlord', 'tenant' or 'managing agent', in which case the relevant duties described in sub-sections 4.5, 4.6 and 4.7, respectively, will also apply.

In particular, the *CAWR* impose specific duties on the employer in respect of asbestos. The table below indicates the employer's responsibilities under the *CAWR*.

Scope/subject	Requirement
Applicability	Applies to all employers and self-employed persons
Prevent/reduce exposure	Prevent or reduce the exposure of employees to asbestos in the workplace to the lowest level reasonably practicable, without relying on respirators
Minimize number	Ensure that the number of employees who are exposed is as low as is reasonably practicable
Immediate response	Take immediate steps to remedy the situation where the amount of asbestos inhaled exceeds the relevant 'control limit'* (this varies according to asbestos type)
Protective clothing	Ensure that adequate and suitable clothing is provided for and worn by employees exposed or liable to be exposed to asbestos, unless the quantity liable to be deposited on clothes is not significant
Emergency measures	Unless the risk assessment shows that because of the quantity of asbestos present, the health risk is only slight, have emergency procedures in place, with information on the emergency arrangements made available to the relevant accident and emergency services (internal and/or external)
Prevent/limit spread	Prevent the spread of asbestos from any place where work is under the employer's control, or, where this is not reasonably practicable, reduce the spread to the lowest level reasonably practicable
Monitor exposure	If the exposure to asbestos is liable to exceed the 'action level'*, monitor the exposure of employees by air monitoring at regular intervals, and whenever a change occurs that may affect that exposure, and keep suitable records

*See Appendix 4 'Regulations' for details.

In essence, the employer must know whether or not his or her workplace contains asbestos, and whether the work activity may expose his or her employees or others to asbestos, and must take appropriate measures to prevent, reduce or control the risk arising to the lowest level reasonably practicable in the circumstances.

4.3 Employer of surveyors

Unless it is a sole trader, a surveying firm will also be an employer, with responsibilities for the safety and welfare of employees (both permanent and temporary staff). In addition, regulations do not make any distinction between a self-employed person and an employer or employee.

For the purposes of this sub-section, only employees who inspect buildings or conduct other services that could expose them to asbestos are considered. The responsibilities associated with risks in the workplace (including offices) apply equally, but are addressed above, under 'Employer' (sub-section 4.2).

Exposure to asbestos while inspecting or administering works

Scope/subject	Requirement
Applicability	Applies to any work involving exposure to any form of asbestos, including sampling and laboratory analysis
Risk assessment	Prior to starting work, assess the likely level of risk and the nature and degree of exposure, and produce a plan to prevent or control this
Training	Employees must be informed and instructed about the risks that they may face, the appropriate control measures and their use of personal protective equipment (PPE) provided by the employer
Equipment	Suitable PPE must be provided and the employer must ensure that it is properly used and maintained (including testing of respirators)
Monitor	The employer must monitor and record exposure where appropriate
Health records	Where exposure to respirable fibres exceeds stated limits ('action levels')*, regular medical surveillance is needed. Records must be kept for at least 40 years

*See Appendix 4, 'Regulations'.

4.4 Employer of unlicensed persons carrying out asbestos works

'Major works' involving asbestos insulation, coating or insulation board or the clearance of contaminated land, may only be carried out or managed by a person or organization holding a licence issued by the Health and Safety Executive (HSE) under the *Asbestos (Licensing) Regulations* 1983.

This guidance note is not intended for employers of licensed asbestos removal contractors, who are expected to be familiar with all the regulations and controls that apply to their specialist work. It is instead designed for the building owner or occupant arranging for his or her employees to carry out general maintenance or building works involving ACMs. These may be either 'minor works',* or involve asbestos cement and other exempt materials, for which an asbestos licence is not needed.

*See Appendix 4 'Regulations' for an explanation of 'minor works' and for other details.

Planned exposure to asbestos during maintenance or minor building works

Scope/subject	Requirement
Applicability	Applies to any work involving exposure to any form of asbestos, including sampling and laboratory analysis
Identify	Determine type of asbestos
Notify*	Inform the enforcing authority
Risk assessment	Prior to works, assess the likely level of risk and the nature and degree of exposure and produce a written plan to prevent or control this
Control exposure	Prevent, or where this is not possible, reduce exposure to the lowest level reasonably practicable, without relying on respirators
Training	Employees must be informed and instructed in the risks that they may face, the appropriate control measures and their use of the PPE provided
Equipment	Provide suitable PPE and ensure that this is properly used and maintained (including testing of respirators)
Monitor	Monitor and record exposure where appropriate
Health records*	Ensure regular medical surveillance is conducted, with records to be kept for at least 40 years
Facilities	Provide suitable washing and changing arrangements
Cleanliness of premises and plant	Ensure that the premises and any equipment used are kept clean during the works and thoroughly cleaned on completion
Movement of waste	Ensure that asbestos waste is properly stored, dispatched and disposed of

*These requirements are only applicable if the level of exposure to respirable fibres exceeds stated limits ('action levels') – see Appendix 4 'Regulations'.

4.5 Landlord

Landlords may also be employers of caretakers, concierges, security staff and in-house maintenance crews who work in their premises, in which case the duties of the 'employer' will also apply (see sub-section 4.2).

In addition, if landlords own non-domestic premises for which they have responsibility for maintenance or repair, or over which they exert control, to any extent, then under the *CAWR* they also have specific duties in respect of the management of ACMs in the parts for which they are responsible. (See sub-section 4.9 'Regulation 4 (1) Dutyholder'.)

The landlord can transfer the responsibility for maintenance and repair and/or the control of premises, and the *CAWR* duties that go with this, to another party. This can be arranged by entering into an appropriate contract appointing a managing agent or a contractor to take on these responsibilities.

In order for an agent or contractor to take on and share legal responsibility, the contract must give them unfettered decision-making powers, as well as the financial control to authorize the necessary expenditure to arrange inspections and to produce and instigate a plan to manage asbestos. (See sub-sections 4.7 'Managing agent' and 4.10 'Delegation'.)

Irrespective of any responsibility for maintenance, repair or control of the premises, the landlord also has a duty to co-operate with other 'dutyholders', including tenants and occupants, as far as is necessary to enable them to comply with their duties under the *CAWR* to manage asbestos. (See sub-section 4.8 'Everyone (duty to co-operate)'.)

4.6 Tenant

Tenants may also be employers, in which case the duties of the 'Employer' outlined in sub-section 4.2 will also apply. In addition, if tenants rent or occupy non-domestic premises for which they have responsibility for maintenance or repair, or over which they exert control, to any extent, then under the *CAWR* they also have specific duties in respect of the management of ACMs in the parts for which they are responsible. (See sub-section 4.9 'Regulation 4 (1) Dutyholder'.)

Irrespective of responsibility for maintenance, repair or control of the premises, the tenant also has a duty to co-operate with other 'dutyholders', including the landlord and other tenants, as far as is necessary to enable them to comply with their duties under the *CAWR* to manage asbestos. (See sub-section 4.8 'Everyone (duty to co-operate)'.)

Similarly, if the tenant is a 'dutyholder', then all other persons, including the landlord, the managing agent, other occupants and neighbours (including those who designed or built the premises, or who have information that would help to locate ACMs or confirm their absence), have a duty to 'co-operate' with the tenant.

These duties are identical to those of the landlord, as set out previously in sub-section 4.5. The extent of responsibility will be determined by the covenants contained in the lease or other form of contract or, in the absence of these, by the actual circumstances on site – i.e., those parts over which the tenant has actual control.

In the final resort, the courts will allocate responsibility as they deem appropriate. However, it is strongly recommended that such issues are resolved and formally agreed with all parties (the other occupants and the landlord) at the earliest opportunity, to avoid recourse to the courts.

It may also be advantageous to arrange for an assessment of ACMs in the entire building, irrespective of control, and to produce a single, comprehensive, management plan. If this course of action is taken, care will need to be taken to allocate responsibility for its safekeeping and updating.

4.7 Managing agent

Very often, as in the case of an absentee landlord, the physical and financial 'control' of premises may lie in the hands of a managing agent or facilities manager.

These people have a significant role, and may be acting as the appointed 'agent' of the landlord, assuming the role of dutyholder and the legal responsibilities to arrange for the management of ACMs in the building. They are also obliged to co-operate with the tenants and other occupants in the fulfilment of their duties. (See sub-section 4.5 'Landlord' above.)

The extent of their control and their ability to finance or instruct the preparation of a 'plan' will depend on their terms of contract, but as a minimum they should inform the landlord of its duties under the *CAWR* and advise as to how these may be complied with. (See sub-section 4.10 'Delegation'.)

4.8 Everyone (duty to co-operate)

Regulation 4 (2) of *CAWR* 2002 (which comes into force on 21 May 2004) gives 'every person' a duty to co-operate with the 'dutyholder' so far as is necessary to enable the 'dutyholder' to comply with his or her duties under this Regulation.

Possible parties

Possible parties include the landlord, tenants, occupants, managing agent, contractors, designers and planning supervisors of non-domestic premises. Under the possible scenarios envisaged by the Approved Code of Practice (ACOP) to the Regulation, possible parties are:

- anyone with relevant information on the presence (or absence) of asbestos; and

- anyone who controls parts of the premises to which access will be necessary to facilitate the management of asbestos (i.e., its inspection, control, removal, treatment or monitoring).

Costs

Co-operation does not extend to paying the whole or even part of the costs associated with the management of the risks of asbestos by the 'dutyholder(s)', who must meet these personally.

Guidance in the ACOP states that architects, surveyors or building contractors who were involved in the construction or maintenance of the building, and who may have information that is relevant, 'would be expected to make this available at a justifiable and reasonable cost'.

Extent of assistance

The duty to co-operate is not subject to any limitation or exclusion. It is not tempered by reasonableness or 'reasonable practicability' and there is therefore an obligation to do whatever is necessary to co-operate with the 'dutyholder'.

For example, a landlord with a lease covenant that, in the event of the default of the tenant, gives the right to enter and carry out works to ensure compliance with statutory provisions, as a last resort would be obliged to pursue this option and claim back the costs as part of the service charges.

Short-lease tenants, licensees or other occupants who control access, but do not have any contractual maintenance liabilities, would be required to permit the landlord access to fulfil his or her duties.

General duty of co-operation

Regulation 11 of the *Management of Health and Safety at Work Regulations* 1999 requires employers who share a workplace to co-operate in order to comply with the relevant statutory provisions.

4.9   Regulation 4 (1) Dutyholder

This Regulation comes into force on 21 May 2004.

The 'dutyholder' responsible for the management of asbestos in non-domestic premises, as set out in Regulation 4(1) of *CAWR* 2002, is every person who, by virtue of a contract or tenancy, has an obligation for the repair or maintenance of those premises, or, in the absence of such, the control of those premises or access thereto or egress therefrom. This includes those persons with any extent of responsibility for the maintenance or control of the whole or part of the premises.

Parties who may be dutyholders include landlords, tenants, occupants, managing agents and managing contractors. Where there is more than one dutyholder, the relative contribution required from each party in order to comply with the statutory duty will be shared according to the nature and extent of the contractual or tenancy repair obligation or the physical control of each.

This Regulation does not apply to 'domestic premises', namely, a private dwelling in which a person lives. However, legal precedents have established that common parts of flats (in housing developments and blocks of flats) are *not* part of a private dwelling. The common parts are classified as 'non-domestic' and Regulation 4 therefore applies to them.

It does not, however, apply to the individual flats or houses or to the kitchens, bathrooms or other rooms within a private residence that are shared by more than one household, or to communal rooms within sheltered accommodation.

Typical examples of common parts are entrance foyers; corridors; lifts, and their enclosures and lobbies; staircases; common toilets; boiler rooms; roof spaces; plant rooms; communal services, risers and ducts; external outhouses; canopies; and gardens and yards. See Appendix 5 for a flow chart to help identify the dutyholder in a variety of types of premises, tenures and modes of occupation and Appendix 8 for a chart showing whether residential premises are likely to be classified as domestic or non-domestic for the purposes of Regulation 4.

For ease of reference, the duties of Regulation 4 are summarized below. For full details, see Appendix 4 'Regulations'.

Subject	Requirement
Co-operate	Take reasonable measures to enable other employers (tenants, other occupants and neighbours) to fulfil their duties
Locate ACMs	Take reasonable steps to locate materials likely to contain asbestos and assess their condition Presume that materials contain asbestos unless there is strong evidence to the contrary
Records	Keep an up-to-date written record of the location, type (where known), form and condition of ACMs
Risk assessment Management plan*	Assess the risk of exposure from known and assumed ACMs Prepare and implement a management plan to control these risks
Review and monitor	Regularly review and monitor the plan to ensure that it is current and is being implemented. Record the findings and actions
Provide information to others	Give necessary information to anyone who is liable to work on or disturb the ACMs, and to the emergency services

*See section 10 for details and advice on the management plan.

4.10 Delegation

As a general point of health and safety law, legal responsibility cannot be delegated. Therefore, if a dutyholder – for example, a landlord with full repairing obligations – employs a managing agent or a contractor to take over this contractual responsibility, both parties will be deemed to be the dutyholders and either or both can be prosecuted for contravention of any of the provisions of the *CAWR*.

For a successful defence, the person charged would have to prove that 'the commission of the offence was due to the act or default of another person, not being one of his employees and that he took all reasonable precautions and exercised all due diligence to avoid the commission of the offence' (ACOP).

In addition, in the event of any criminal proceedings, the defendant must, within a specified period of time, serve a written notice identifying (or assisting in the identification of) the dutyholder(s) responsible. In the case of England and Wales, this is seven clear days before the court hearing, and in Scotland, the same period before the intermediate diet, or the first diet, where the proceedings are summary or solemn, respectively.

Of course, in this case, the landlord would need to have assured him or herself that the appointed party was competent and properly resourced to fulfil the duty. The same would apply if a tenant appointed another party to undertake inspections of the premises and to identify materials containing asbestos on his or her behalf. (See Appendix 7 for further details and advice.)

To avoid any confusion or misunderstanding about the extent of their duties, a managing agent or contractor who is unable or unwilling to take on such onerous statutory responsibilities should ensure that their contract specifically excludes any 'duty to manage asbestos' responsibilities.

If a person intends to employ others on the basis that they are not responsible for their health and safety, then legal advice should be sought before proceeding.

Part 5
Services offered by surveyors

5.1 General

Surveyors provide a wide range of services, and the type, extent and scope of inspection and investigation required for each varies accordingly. For example, that required for a 'valuation' purpose is very different from that needed for a 'building survey'.

In addition, the scope of inspection will vary according to the specific requirements, as set out in the terms of engagement agreed in each case. These may range from a general overview of major significant matters – for example, noting the presence of insect-infestation of timber, or its decay – as part of a wider, all-encompassing survey of a building, to a much more detailed analysis of this particular issue, for example, to establish the detail required for the preparation of a specification for remedial treatment or management of the timber problems.

In both cases, the element concerned is the same – timber components – but the depth of detail required is very different. It is this that dictates the nature, duration and extent of the inspection or investigation required.

Thus, there are elements that are only incidental, and which form part of a wider picture, and those that are specific and are confined solely to a particular subject or to a limited number of subjects.

In both cases the degree of knowledge, experience and training required by the surveyor may also be very different, and this will be reflected in the performance of the services that can be reasonably expected.

The same consideration applies equally to asbestos. The expectations of an inspector commissioned for a project that is only and specifically concerned with asbestos are different from those reasonably required of the surveyor obliged to comment on asbestos incidentally in the generality of numerous other matters.

For the purposes of this guidance note, and to differentiate between these two extreme levels of service, it is presumed that the 'incidental' consideration refers to asbestos as only one part of a building survey or valuation, and is undertaken by a surveyor or valuer, and the latter 'specific' type to an 'asbestos inspection' conducted by an 'asbestos inspector'.

The term 'asbestos inspector' is not widely used at present, but it is a useful way of distinguishing the levels of expertise, knowledge and experience required for this specialism from the general knowledge of all matters affecting buildings that is required of the surveyor or valuer. (See sub-section 1.6 'Asbestos survey or inspection…surveyor or inspector?' for discussion of the differences between these terms.) It is also possible that future personal certification measures that are under contemplation will use this designation. (See Appendix 7 for further details.)

The following sub-sections address the possible implications of asbestos for the various services provided by RICS members, under the headings of these services. The sub-sections contain examples of typical situations, which are used to illustrate the relevance for a particular service. However, the situations are not unique to any one service, and often are equally appropriate for other types of service. It is important, therefore, that service areas are not viewed in isolation, but are read in conjunction with the others and with the general advice given in the preceding sections of this guidance note.

5.2   Auctions and tenders

Auctioneers should be alert to the possibility that buildings, land and other items which they may sell or auction on behalf of others may contain asbestos, and should be aware of the risks associated with its inadvertent disturbance or uncontrolled removal.

External lagging or insulation should be obvious; however, old industrial plant and machinery, particularly boilers, pipes and calorifiers, may well contain asbestos and components that could be 'hidden' in the form of gaskets, washers or even 'flash arresters' in electrical fuse boxes.

Plant and equipment containing asbestos can only be sold if either:

- it is fixed within the building and is being sold as part of that building in situ; or

- it is part of a vehicle or mobile plant, for example, a tractor or crane.

Under Regulation 4 of the *Control of Asbestos at Work Regulations (CAWR)*, the seller must inform the buyer of the presence or reasonably presumed presence of asbestos, and its condition, as he or she is liable to disturb it.

Plant and equipment is sometimes sold with a condition attached that it is removed and dismantled by the purchaser. There is a danger that without specific sale conditions, this work may be carried out in a hurry, illegally, by untrained, unskilled workers, without the use of appropriate equipment or controls. Such workers may be interested only in preserving the plant with value (possibly only scrap value) and this could result in the careless stripping and discarding of unwanted insulation and cladding.

Not only is this illegal, but the uncontrolled removal of plant, equipment and components containing asbestos can result in the contamination of adjacent plant, surfaces, services or even entire buildings. The disruption involved in the specialized decontamination subsequently needed may far exceed the proceeds of the sale of the plant. Depending on the terms of engagement and the particular circumstances, including the degree to which he or she has control over the removal operations, the auctioneer may be liable for this contamination. In any case, in the example quoted above, the consequences could seriously damage professional reputations and careers.

Just as it is now common practice to refuse to sell electrical goods without first obtaining a test certificate, it is prudent to refuse to sell products that are likely to contain significant amounts of asbestos without considering the risk to:

- members of the general public attending the sale;

- auctioneering staff (inspecting, valuing, displaying and delivering the products);

- potential purchasers examining the items;

- purchasers or their contractors when removing or dismantling items; and

- other premises or products, through contamination.

(See also sub-section 5.6 'Fine arts and antiques'.)

5.3   Reinstatement cost assessments

Surveyors undertaking reinstatement cost assessments should be aware of the following considerations in respect of damage (caused by fire or otherwise) to buildings containing asbestos components and materials.

If it is necessary to replace damaged components that contain asbestos, for example, corrugated asbestos cement sheeting to the walls and roofs of industrial buildings, surveyors should note that since 1999 it has been illegal to purchase and use material containing any form of asbestos. (There are a few exceptions, such as specialist filters used in laboratories and scientific institutions, but generally these are outside of the scope of the majority of buildings in which surveyors will be in regular contact.)

Asbestos materials have numerous qualities and the original components may have been providing a number of different and separate functions, such as fire protection and thermal and sound insulation. It may be difficult to find a suitable replacement material with all of the required qualities, and a combination of various materials or components may therefore be necessary.

However, substitute materials may lack the strength and qualities of the original components. For example, some mineral fibre (asbestos-free) roof cladding is up to 15% weaker than the original asbestos. The supporting structure may need to be adapted or redesigned to account for this.

There may also be difficulty in finding compatible replacement materials, such as matching corrugated cladding or sheeting, although manufacturers have developed replacements for the more common types.

As well as considerations of replacement, there is the risk of the potential contamination of neighbouring buildings and land during a fire or explosion. This is particularly significant with asbestos cement products, which have been known to explode in serious fires, due to the intense temperatures generated, with debris spread over a wide area. The costs of cleaning up contamination, including the charges of the emergency services, can be substantial.

The situation may be far worse in the event of a general explosion. A blast could spread asbestos products, which are usually relatively lightweight, over a large area, in the form of contaminating rubble and debris. This may then need to be cleaned or disposed of as contaminated waste, at considerably greater expense than untainted material that could perhaps have been recycled.

Because of their relative cheapness and durability, asbestos materials were extensively used in low-cost industrial and agricultural buildings in the past. The costs of treating, decontaminating and clearing up asbestos after a fire or other disaster in such premises could represent a significant proportion of the cost of reinstating the property.

5.4 Historic and listed buildings

It is highly unlikely that buildings constructed prior to 1800 will have had asbestos materials incorporated as part of their original design. It is, however, possible that asbestos-containing materials (ACMs) may have been added during their lifetime, as part of structural alterations, general refurbishment or the replacement of services installations, in the form of fire protection or thermal or sound insulation.

Consequently, unless it can be guaranteed that the historic building is entirely original, it cannot be assumed that it is free of asbestos, and care should be taken when inspecting, repairing, altering or demolishing it.

5.5 Agricultural and rural undertakings

Because asbestos was a cheap, durable and readily available material, it was frequently used in the past in low-value buildings such as barns, outhouses and other buildings associated with farms and other rural businesses. The most common form was asbestos cement products, owing to the water-resistant quality of this form and its ability to be moulded into various profiles, as wall cladding, roof sheeting, pipes, gutters and other rainwater goods, cisterns and tanks.

Other forms of asbestos may have been incorporated into these buildings as well, perhaps as internal linings, pipe insulation or lagging. However, these are not such common forms, as they were not suitable for the main use for this economic sector, which was to serve as a cheap, waterproof external enclosure that could be quickly and easily erected, altered and dismantled.

For these buildings, while the material is intact and in good condition, the asbestos fibres are safe, as they are bound into and sealed by the cementitious matrix. The problem arises when this matrix is disturbed, whether by impact damage, fire or natural degradation due to age and exposure, with the latter accelerated by moss and lichen growth.

Very often, the enclosures are not used for human occupation, but as grain stores or animal shelters. In addition, they may be well ventilated, dispersing and diluting any airborne asbestos fibres. Thus the health risks to workers are generally minimal, except when carrying out works to the asbestos materials, during maintenance, repair, replacement, alteration or demolition.

In the case of asbestos cement materials, there is no requirement for licensed asbestos contractors to be used; however, the persons carrying out such work must be competent (must know the risks and be properly equipped) and certain basic health and safety precautions must be taken in every case.

Because of the fragile nature of asbestos cement sheets, which become more brittle with age, when working on roofs or at height the risk of injury from falls is often far greater and more immediate than that of contracting asbestos-related disease. The control measures should therefore be commensurate with the risks arising.

The Rural Design and Building Association (RDBA) has published a guide entitled *Working with Asbestos Cement Products.*

(See also sub-section 5.3 'Reinstatement cost assessments' and 5.10 'Dilapidations, leases and agricultural tenancies'.)

5.6   Fine arts and antiques

RICS members involved in such artistic and aesthetic pursuits as those associated with fine arts and antiques may consider that asbestos is irrelevant to their activities – and in the generality of the service they provide, this is often the case.

There are, however, circumstances when the ability to recognize materials that might contain asbestos, and an awareness of the health risks arising, may be absolutely essential to their own well-being or that of others.

For example, it may be necessary to inspect or search for a work of art in an attic, undercroft, store or cellar that could be heavily contaminated by deteriorating or damaged ACMs, perhaps in the form of pipe or boiler lagging. In addition, the physical removal or movement of an item for sale may entail the disturbance of an adjacent material containing asbestos, as for instance, in the case of a need to enlarge an opening in a partition or ceiling to remove a work of art, or to dismantle ornate panelling, thereby damaging or leaving exposed asbestos materials.

In order to meet the responsibilities as an employer, a basic knowledge of asbestos, of its likely locations and of when to employ or seek specialist advice is essential.

(See also sub-section 5.2 'Auctions and tenders'.)

5.7   Quantity surveyor/cost consultant

The role of the quantity surveyor varies considerably, and legal responsibility for asbestos, like most other health and safety hazards, depends to a large degree on the surveyor's autonomy with regard to design and contractual decisions.

Often, a quantity surveyor will only be interpreting, refining or devising a way of measuring the requirements of other designers, such as architects, service engineers or even the client. On other occasions, they may be choosing materials, dictating methods and sequences of working, or selecting and appointing contractors or other specialists, perhaps in the role of project manager or as the employer's agent. In both cases, they take on the role and responsibilities of a designer and, in the latter, possibly those of the client as well.

The circumstances in which a knowledge and understanding of asbestos would be relevant, and the issues to be considered, include the following:

- personal safety, when visiting properties containing unsealed or damaged asbestos for the purposes of estimation, measurement or contract administration;

- the financial implications of works involving the disturbance or removal of asbestos;

- programming issues;

- the preparation and review of tenders – i.e., the level of competence required of contractors and other specialists, and the factors to be taken into account in their assessment and appointment;

- assessments of reinstatement costs, and works to fire-damaged premises; and

- advice on tax relief for the costs of remediation of contamination (see sub-section 5.12 'Tax advice').

Since 1999 it has been illegal to specify the use of materials that contain any form of asbestos, except in very few and specialist circumstances, such as for highly specialist filters, for which a suitable alternative is not yet available.

5.8   Valuations

Valuations are required for various purposes, and the Standards in the RICS *Appraisal and Valuation Standards* (the *Red Book*) will apply.

PS 4.1 requires the valuer to make such inspection and investigations as are needed to produce a valuation that is professionally adequate for its purpose. Valuers are not usually expected to undertake detailed investigations into or reach definitive conclusions on all matters listed in the practice statement, which includes asbestos, but should make these limitations clear to the client when agreeing the terms of engagement. As these matters can rarely be disregarded completely, the discovery of on-site factors that may affect the valuation should be drawn to the attention of the client before a report is issued.

Although PS 4 applies to all valuations, there are Standards that apply specifically in the UK:

- UK Appendix 3.1, which applies to commercial secured lending, does not mention asbestos, but contains a requirement that the valuer shall comment on any material disrepair and any assumptions about future repairs;

- UK Appendix 3.2 applies to residential property mortgages. It states that the valuer need not make enquiries regarding contamination or other environmental hazards, may assume that no deleterious or hazardous materials have been used in the construction of the property, and is under no obligation to verify such an assumption. However, if a problem is suspected, the valuer should recommend further investigation. In addition to advising on the security of the loan, the information on the condition of the property is also likely to be used by the purchaser, for example, in determining the extent of the works needed to put the property into good repair;

- UK PS 4.1 requires a surveyor providing the RICS Homebuyer Survey and Valuation (HSV) service to comply with the material published by RICS. The published material also confirms that the surveyor will not research the presence of harmful substances, but states that if their presence is suspected, advice will be given on the action to be taken.

- UK Guidance Note 1 also refers, briefly, to asbestos, but refers the valuer to this guidance note.

Asbestos is a common hazard (irrespective of the actual risk) and, depending on the circumstances, its presence or condition can materially affect value. As a result, it cannot be ignored – but to what extent should it be commented on?

While asbestos may not constitute an active risk to health in situ, the costs of its treatment or removal can be considerable. In addition, the cost of building alterations or maintenance that involve the disturbance of asbestos will be increased. Accordingly, knowledge that asbestos is present can adversely affect the value of an interest in a property. Knowledge of its presence could dissuade potential purchasers, thus limiting the demand for a property and lowering its market price.

The actual effect on the value of the premises as an investment is dependent upon the circumstances in each case, and will vary according to the amount and type of asbestos, its condition and location, the ease with which it can be treated, or removed if necessary, and the use, size and type of the building. The presence of a small amount of asbestos in good condition should have only a negligible, or no effect on the value of a property.

When providing a valuation, the asbestos element is one of a number of matters to be considered, and the surveyor cannot be expected to devote the same degree of time, attention and detail to this single topic as in a specific, stand-alone 'asbestos inspection'. (See sub-section 1.6 for a definition of this term.)

It is comparatively rare for asbestos to be an overriding consideration in a valuation. If the valuer believes this to be the case, it would be appropriate either to decline to provide a valuation until a specialist report has been obtained, or to agree that the valuation be qualified with a clear and unequivocal statement that asbestos is present and that the valuation is subject to the findings of a specialist report on the material. It is more usual for the presence of asbestos to be only one of a number of matters to be considered, any one of which may affect the valuation.

With regard to asbestos, the following enquiries are recommended during a valuation.

- Reasonable enquiries should be made as to what existing information on asbestos is available, in the form of previous survey reports, the asbestos register and, from 21 May 2004, the 'asbestos management plan'.

  If not provided in the form of separate documentation, this information might be contained in a health and safety file prepared in accordance with the *Construction (Design and Management) Regulations* (*CDM*) 1994.

  Currently, neither the *CDM* Regulations nor the *CAWR* apply to work for 'domestic clients' or to 'domestic properties', and thus in the majority of cases these statutory documents will not be available to surveyors inspecting domestic residential buildings. (These exclusions are not total in all circumstances, and reference should be made to the appropriate regulations for precise details if required.)

  The introduction of the concept of the management plan means that the significance of existing registers will decline over time, until they are integrated into this new document. For guidance on assessing the completeness or competence of asbestos management plans, see section 10.

  Like other important documents such as fire and building regulation certificates, the availability or absence of these documents should be noted, including any reasonable inference that may be drawn. Where appropriate, a recommendation should be made for the client's legal advisor formally to check their existence and/or validity, as appropriate.

- Reasonable enquires should be made to determine what, if any, changes have subsequently been made that could affect the usefulness of these documents. For example, ACMs noted as presenting an immediate risk may have been subsequently repaired, treated or even removed.

  It is reasonable for the valuer to rely on the occupant's assurance that the information is current and accurate, where there is no obvious physical evidence to the contrary noted during the visit. In the case of management plans, reference to the dates of the latest review and monitoring will assist in this.

- Inspection of the premises should be to the extent dictated by the commission, and as agreed with and confirmed to the client. If, during the inspection, the surveyor notes asbestos which in his or her opinion constitutes an immediate and serious risk to health, the person* in control of the premises should be informed as soon as possible, and be given advice on the emergency measures required. This should be confirmed in the report, together with details of the person informed, and a note of the date and time, where appropriate. (See sub-section 3.2 'Statutory responsibility'.)

*This may include the householder or, in the case of vacant properties, the estate agent or property manager.

Any assumptions made should be recorded. Where appropriate, the record should clarify the basis of these, for example, from details obtained from reports, documents or other information supplied by others, or confirmed personally during the inspection.

5.9 Building surveys

The term 'building surveys' is often used generically, and may be interpreted in a variety of ways by different parties. It is therefore essential that the context, including scope and limitations, is made clear from the outset, when agreeing the terms of engagement for a particular purpose.

A 'building survey' is very different to an 'asbestos inspection', in which the sole aim is to identify and report on materials containing asbestos. For the latter, the scope and depth of investigation and comment on asbestos is much greater. (See section 8 'Asbestos inspection'.)

RICS has published guidance notes to clarify these issues for building surveys of particular types of premises, and for particular purposes. These guidance notes include *Building Surveys and Inspections of Commercial and Industrial Property, Building Surveys of Residential Property* and *Contamination and Its Implications for Chartered Surveyors*. The RICS *Homebuyers Survey and Valuation Licensing Scheme* may also be of help. RICS members are of course required to be familiar with the Institution's official minimum requirements for each of the services that they provide.

The services provided by members vary in terms of the way in which asbestos is addressed, both specifically and by implication as a deleterious or significant material. The services also have subtle differences in terms of the scale of investigation it is usually deemed appropriate to undertake.

It is not within the scope of this document to consider these differences in depth. Moreover, at the time of publication, a number of RICS publications were under review, and the relevant parts may therefore be changed.

The matters of significance, which may vary in each case, are:

● applicability (for example, to England and Wales, or Scotland or the UK as a whole);

● specific references to asbestos, and quoted examples;

● indirect references to asbestos, for instance, as a contaminant, a deleterious material or a pollutant;

● tests and the employment of specialists;

● areas or elements to be inspected or not (such as roof spaces, floor voids, and the common parts of leasehold property);

● services installations;

● land contamination; and

● sources of information (for example, the asbestos register, management plan or health and safety file).

5.10 Dilapidations, leases and agricultural tenancies

General

A schedule of dilapidations is a document that records breaches of contract in relation to building fabric and service installations. A breach must exist before a tenant has a liability to pay damages to the landlord, or vice versa.

Although many buildings have asbestos in their fabric or service installations, it is currently extremely rare for asbestos to be specifically addressed in leases or licences. It is therefore necessary to consider it in the generality of other building components that are usually less harmful and more easily repaired or replaced.

Repairing covenants

In the case of any repairing covenant, there must be some form of deterioration since the contract was entered into for a breach of covenant to exist. The mere presence of asbestos will not constitute a breach of covenant, and there is no obligation to repair or remove it unless it has been damaged or has deteriorated.

With regard to deterioration, it is established law that in order for there to be a breach of covenant, the deterioration must be greater than that usually expected for the type of building over the term of the lease – that is, below the standard contemplated by the covenant. This may be relevant when considering external asbestos cement cladding or roofing, which degrades naturally as a combined result of age and exposure.

Where actionable damage has occurred, the landlord, tenant and their professional advisors should be aware of the potential additional complications and costs that ACMs can create. For example, the new use of any material containing blue, brown or white asbestos is illegal, with these types having been banned in 1985, 1992 and 1999, respectively. It will not therefore be possible to replace damaged ACMs like for like, and a suitable substitute material or component will be required.

In some cases, it may not be possible to replace all of the properties of the ACM using a single substitute material or component; instead, a combination of materials or components may be required, with the supporting structure adapted to suit this.

Unless the works are only 'minor', then they must be carried out by a licensed asbestos contractor. (See Appendix 4 'Regulations' for a detailed explanation of the *Asbestos (Licensing) Regulations* 1983.) Among other cases, the definition of 'minor works' includes works carried out by the occupants of the premises themselves (that is, the occupants' in-house maintenance team). However, the members of this team must be adequately trained and equipped for such activities.

In the case of asbestos cement materials, there is no requirement for licensed asbestos contractors to be used. Nevertheless, the persons carrying out the work must be competent and properly resourced.

Compliance with statutory regulations

Many leases include a covenant for compliance with current statutory requirements.

Failure to comply with the *CAWR* is of course a criminal offence. However, the inclusion of a parallel contractual requirement also provides the opportunity for civil action, to preserve the party's interest or obtain financial recompense in the event of any breach of a lease covenant.

There are two key areas of concern with regard to asbestos: first, ensuring that works involving ACMs are properly carried out by competent persons, using appropriate controls; and second, ensuring that any asbestos that remains is properly managed.

In the former case, the danger is that an incompetent person may contaminate parts or even the whole of a building and/or neighbouring premises, with all of the risks to health and additional financial burden that this could entail. An example of this would be the contamination of common service risers in a multi-let office building during the refurbishment of a tenanted area in the top storey.

In the case of the statutory duty to manage asbestos, Regulation 4 of the *CAWR* only applies to 'non-domestic premises'. However, this also includes the common parts of housing developments and blocks of flats. (See Appendix 4 'Regulations' for details of the Regulation, including how to identify the dutyholder(s), and Appendix 8 for a chart showing whether residential premises are likely to be classified as domestic or non-domestic for the purposes of Regulation 4.)

Essentially, the Regulation requires the dutyholder – who could be the landlord or the tenant, depending on the repairing obligations in the lease or licence, or the actual controller of the premises – to make an assessment of any ACMs in the parts of the demise for which they are responsible and to produce and implement a plan to manage these. The arrangements of the plan must be periodically reviewed and the condition of any remaining ACMs monitored.

The significance of the plan is that it demonstrates to others, including potential purchasers or lessees, that the content of asbestos has been identified – or at the least, 'presumed' – and that they are therefore not likely to come across asbestos unexpectedly and be faced with unplanned expenditure in the future.

This has to be tempered with the appreciation that often, without substantial dismantling and disturbance that is usually unacceptable to the occupants or owner, it is not possible to discover all ACMs, for example, those hidden in linings to service ducts, chimneys and in linings behind shop fittings. It is thus common for there to be grey areas, where the presence of asbestos or otherwise cannot reasonably be confirmed, perhaps until dismantling or demolition takes place.

The management plan is an important legal document and may be expensive to create in the event of its earlier omission, or to reproduce if lost.

The plan is owned by the dutyholder who prepared it. However, Regulation 4(9)(c)(i) requires the owner or leaseholder to ensure that relevant information is passed on to the new occupier. Guidance in the Approved Code of Practice (ACOP) to the Regulation states that any costs associated with making the plan available are to be justifiable and reasonable. It is assumed that costs will be limited to the costs of retrieval and printing and not represent a proportion or the whole of the value of the plan to the new incumbent.

Co-operation

Regulation 4(2) of the *CAWR* requires 'every person' (including both the landlord and tenant(s)) to co-operate to enable the dutyholders to comply with their duties (i.e., to assess the ACMs and to produce and maintain a management plan). The duty is without limitation. In some cases, but only where the covenants of the lease permit, it may be deemed reasonable – or indeed essential – for the landlord or head lessee to arrange for the works to be undertaken themselves, and to recoup the costs as part of their service charges.

The various dutyholders of different parts of the premises will need to know the details of each other's management plans, to ensure that all areas of the building are properly addressed and that there is no conflict or any unnecessary duplication.

Removal

There is no statutory requirement to insist on the removal or treatment of asbestos. The government line is to encourage ACMs that are intact, in good condition and unlikely to release respirable fibres, to be left alone.

Removal of asbestos is the final solution and such a decision must not be taken lightly. If removal is subsequently deemed by the enforcing authority to have been inappropriate, there is the possibility that the instigator could be prosecuted for exposing people, including the protected asbestos strippers, to unnecessary risk to health.

In some circumstances, the asbestos may have been providing a number of different functions, for example, physical enclosure, fire protection and thermal insulation. In order to restore these properties, as required by the lease, it may be necessary to provide a combination of components that are more complex and expensive than the original ACM. There may thus be an element of 'improvement', the apportionment of the costs of which will depend on the circumstances in each case.

New leases, licences and agricultural tenancies

The repairing and maintenance covenants of a lease, licence or agricultural tenancy agreement will determine the identity of the 'dutyholder(s)' responsible for the management of asbestos in the various parts of a building. When new contracts are agreed, in order to avoid any ambiguity, care should be taken to include reference to the *CAWR* and clearly to identify and allocate responsibility to the dutyholders for all the various parts of the premises.

The lease, licence or agricultural tenancy agreement should include a requirement for seeking the landlord's permission for any works that may disturb known or suspected ACMs, and for updating the management plan accordingly. It would be useful to insist that a copy of the current asbestos management plan is given to the landlord as proof of compliance with Regulation 4 of the *CAWR* and to include a right of entry in the event of the default of the tenant, and to prepare or update the plan on their behalf and at their expense.

The costs of periodically monitoring any remaining asbestos and of reviewing the plan(s) should generally be minimal. However, it would be prudent to consider whether the financial responsibility for the landlord's areas should form part of the service charges.

5.11 Contract administration/project management

The implications of asbestos for contract administrators and project managers are similar to, and overlap with, those for quantity surveyors and cost consultants (see section 5.7). In summary, the main concerns are:

- the competence and resources of specialist consultants and contractors;

- time and programming issues (taking account of the lengthy preparation and clearance operations often involved in asbestos removal or treatment);

- the exclusion of unauthorized persons from works in asbestos areas;

- methods of monitoring the safety and quality of works;

- the certification of removal or clearance prior to the re-occupation of stripped areas;

- proof of the proper disposal of contaminated waste; and

- preliminary items (for example, power, water and space for any decontamination facilities required; the location of sealed skips for contaminated waste; and the siting of exhaust outlets for air extraction hoses).

As a point of principle, investigations to determine the presence of ACMs should be carried out as early as possible, in order to protect persons who may need to visit the premises in advance of the main works, such as engineers, surveyors or even unauthorized persons.

The inspection should be a Type 3 'Full access sampling and identification survey (Pre-demolition/major refurbishment survey)' (for details of this see sub-section 8.3 'Types of survey'). It is not necessary for the asbestos report to contain information or advice on how to manage the material, or to include a risk assessment, if it is intended that all of the ACMs will be removed as part of the proposed building works.

All available relevant details of known ACMs should be included in the tender documents. 'Contractor's risk' clauses such as 'remove all ACMs encountered' are contrary to the *CDM* Regulations. Where the full extent of ACMs cannot reasonably be determined in advance, there must be a facility, such as an agreed schedule of rates, for the contractor to be paid for removing or treating these materials.

If detailed methods of asbestos removal are dictated, then there is the possibility that the author may be deemed to be 'managing or supervising asbestos removal works', rather than simply administering the contract. In this case, under the *Asbestos (Licensing) Regulations* 1983, an 'asbestos removal licence' is required.

The specification/bill of quantities and/or health and safety plan should always include instructions on what to do if any additional suspected ACMs are discovered during the term of the contract. Such instructions may include stopping work that might disturb the materials and reporting the discovery to the person in charge of the site, before seeking further instructions.

5.12   Tax advice

RICS members should be aware that tax relief is currently available under certain circumstances in respect of works that are carried out in relation to asbestos.

As introduced by Schedule 22 to the *Finance Act* 2001, a 150% tax deduction is available to companies incurring expenditure on 'land remediation'. 'Land' also includes buildings and 'contamination' can include asbestos. The company must have acquired the land in its contaminated condition and both developers and investors may qualify.

The tax relief is received as an allowable deduction in computing, for tax purposes, the profits of the company's Schedule A business or trade. As such, it is claimed via the company's self-assessment tax return. As with many forms of tax relief, a number of conditions must be met and technicalities overcome before the benefits can be obtained.

Tax matters can be extremely complex and advice should therefore be sought from an expert in these matters.

Part 6
Limitations and exclusions

6.1   General

Legal advisors to RICS suggest that the only way in which a surveyor can avoid the need for any mention of asbestos during an inspection of a property is if this is a specific requirement of, or has been agreed with, the client. This must always be on the understanding that the client has either already appointed, or will appoint, another party, perhaps a specialist (asbestos inspector) to conduct an asbestos inspection and provide the necessary specialist advice. Such an agreement should be confirmed in writing with the client in the confirmation of the conditions of engagement, so that both parties are fully aware of the fact that asbestos will not be mentioned in the final report.

Irrespective of this, however, the surveyor has overriding obligations imposed by the *Health and Safety at Work etc. Act* 1974 and associated general regulations and is under a general duty to warn of potential danger (see section 3 'Surveyors' responsibilities'). Consequently, where, during the course of an inspection, a surveyor identifies suspected or actual asbestos, or associated contamination, which in his or her opinion presents an imminent actual or potential risk to the health of the occupants, visitors or the general public, he or she should report it, irrespective of the conditions of engagement that apply and the assumptions that have been agreed, on a duty to warn basis.

Where appropriate, this action should include the provision of advice on the emergency short-term measures to be taken, which in extreme circumstances could include evacuation of the contaminated area and prevention of unauthorized entry. It is anticipated that this will only happen in rare circumstances.

Where the surveyor is, by virtue of his or her lack of knowledge, training or experience, not capable of providing the necessary service or advice, then the client must be informed of this. However, the client must not be coerced into agreeing to an exclusion relating to asbestos. Surveyors should remember that a greater duty of care is required when dealing with the domestic client (the man or woman in the street), where it must be assumed that the degree of reliance on the advice and guidance of the surveyor will be greater than that of the informed and experienced client.

6.2   Excluding liability

Personal injury

As a point of law, surveyors cannot wholly exclude liability for personal injuries in respect of any report issued by them to their contracted client, albeit they may not be insured for it. In addition, a surveyor is not able to disclaim liability to individuals, including third parties with whom he or she has no contractual relationship, for personal injuries that arise as a result of his or her negligence in failing to identify asbestos.

The three criteria essential for establishing liability are the following:

- there must be a duty of care owed to the injured party;

- there must be proof that the duty has not been fulfilled; and

- there must be a causal link between the breach of the duty of care and the injury or damage caused.

Damage other than personal injuries or death

Section 2 'Negligence Liability', of the *Unfair Contract Terms Act* 1977 (*UCTA*), states that:

> 'in the case of other loss or damage, a person cannot so exclude or restrict his liability for negligence except in so far as the contract term or notice satisfies the requirement of reasonableness.
>
> Where a contract term or notice purports to exclude or restrict liability for negligence a person's agreement to or awareness of it is not of itself to be taken as indicating his voluntary acceptance of any risk.'

The initial consideration will be to establish whether the exclusion clause was incorporated in the surveyor's terms and conditions of engagement. In order to do this, any standard terms of engagement must be made known to the client and agreed upon at the time when instructions are received. If they are not, there is a real risk that they will be of no contractual effect.

The various guidance notes issued by RICS in respect of inspections of both residential and commercial and industrial property require that the extent to which the inspection will be limited, and any caveats and/or assumptions, must be agreed when taking instructions and confirmed prior to undertaking the survey. In the case of a residential survey, the Model Conditions of Engagement prescribe the scope of the areas and elements to be inspected, and these can only be varied by specific agreement with the client.

There is a requirement, where appropriate, to explain the implications of any changes to the client, with the onus being on the surveyor to provide the explanation. These changes will include those caveats specifically required by the surveyor's professional indemnity insurers, and should also include any specific exclusions on the professional indemnity insurance (PII) policy. The RICS *Building Surveys of Residential Property* guidance note states that the actual wording required by the insurance policy must be used verbatim.

In respect of liabilities arising in contract between contracting parties, where one of them deals as a consumer or on the other's written standard terms of business, the guidance note states the following:

'As against that party, the other cannot by reference to any contract term:

a) when himself in breach of contract, exclude or restrict any liability* of his in respect of the breach; or

b) claim to be entitled:

   i) to render a contractual performance substantially different from that which was reasonably expected of him, or

   ii) in respect of the whole or any part of his contractual obligation, to render no performance at all.

except in so far as (in any of the cases mentioned above in this subsection) the contract term satisfies the requirement of reasonableness.'

UCTA 1977 states two sets of circumstances in which the contracting parties may be able to vary the performance from that specifically stated in the contract. These are where the client is dealing as a 'consumer' with the surveyor, and where the client is not a consumer, but deals with the surveyor on the surveyor's written standard terms of business.

Dealing as a consumer means, in practice, that the client has consulted the surveyor for the purpose of the client's personal affairs and not in the course of a business. A private individual wanting a house for his or her own occupation would be 'dealing as a consumer', but an investment company having a private residential property valued for the purpose of its business could not claim to be dealing as a consumer with the surveyor.

In both circumstances, this is subject to the requirement of the 'reasonableness' of the term, namely that it should be fair and reasonable to be included, having regard to all the circumstances that the parties knew, or ought to have known, at the time they made their contract.

In the case of limitations and exclusions, there are two main conditions for reasonableness to apply: that both parties should have been aware first, of the extent and second, of the implications, prior to the provision of the service.

UCTA 1977 expressly excludes contracts of insurance from its effects, and in any case does not apply to these, because the insurance will be taken out in a business or professional capacity, rather than in a 'consumer' or personal capacity. As a result, this legislation, which governs the use of disclaimers in English law, does not affect the position vis-à-vis the professional indemnity insurers and their surveyor clients. It is therefore open to insurers to exclude from cover all liabilities, including personal injuries, that occur as a result of negligent acts, errors and omissions arising from surveys or other services relating to asbestos. (See section 7 on 'Insurance' for the implications of this.)

---

*For the surveyor there are two types of liability to be considered, in respect of negligence and of liability arising in contract.

Part 7
Insurance

There are three main types of insurance cover that a surveyor might require: employer's liability (only applicable if the surveyor is an employer); public liability; and professional indemnity insurance.

Insurance is of course necessary for any service provided by a surveyor, but it has particular significance in view of the special health and economic risks associated with asbestos-related services.

It is neither necessary nor appropriate for this guidance note to address the issues of employer's liability and public liability insurance. Surveyors are advised to contact the RICS insurance manager, or their broker, if details on either of these forms of insurance are required.

So far as professional indemnity insurance is concerned, the significance of this for the asbestos-related services that are currently provided by surveyors may not be fully appreciated.

In particular, confusion may have arisen over the different interpretations of the terms 'pollution' and 'contamination' and over what is considered to be a 'pollutant' or a 'contaminant'. To many, these terms are associated with, and confined to, contaminated land. Surveyors should be aware of the extent to which, depending upon their skills, knowledge and so on, they are permitted to give advice or otherwise be involved in such matters.

What may not be clear is that the insurance industry generally considers asbestos to be a contaminant, a pollutant, or both, in relation to buildings as well as land (the latter in the context of landfill or brown-field sites, for example).

In order to understand the concern that the insurance industry has with asbestos, it should be appreciated that problems arising from environmental damage and, in particular, those that are asbestos-related, have been and continue to be a major source of claims. Such claims have been measured in billions of dollars. It was claims of this nature that nearly resulted in the collapse of the Lloyds of London insurance market in the 1990s.

The claims to date have mainly originated in the US, but many of the affected underwriters are British syndicates. There is also a tendency for the British markets to mirror those across the Atlantic. In addition, fears have been aggravated by the increasing litigiousness of society as a whole.

Early in 1992, as a consequence of claims arising from environmental damage notified in the US, insurers and underwriters in the British market became alarmed at the extent of the possible exposure to environmental losses. This was aggravated by uncertainties concerning ultimate responsibility for remediation, a lack of environmental standards and a possibility of further legislation by the European Commission. As a result, the wording of most professional indemnity insurance policies was amended to exclude liability for losses arising from pollution or contamination.

In view of the 'claims made' nature of professional indemnity insurance, the effect of this was immediate, and applied to claims notified subsequently, notwithstanding that the work itself may have been performed before the removal of cover.

Since 1 January 1986, RICS has required its members to maintain professional indemnity insurance for claims arising from both breaches of contract and professional duties. The Institution sets down minimum requirements for the level of indemnity, maximum limits of uninsured excess and requirements for the quality of cover. RICS prescribes a policy wording which sets a minimum standard for RICS-listed insurers.

A pollution exclusion clause was incorporated in RICS policy wording on 1 January 1994 and became effective for renewals after that date. The definition of pollution is noted below.

> 'For the purposes of this exclusion, pollution shall mean pollution or contamination by naturally occurring or man-made substances, forces or organisms or any combination of them whether permanent or transitory and however occurring.'

The effect of this pollution and contamination exclusion is wide. By way of explanation, there follows a consideration of the salient points. Surveyors must of course refer to the details of their own specific policies, which may be very different.

The 'services' covered by the insurance are those 'normally undertaken by members of RICS or as otherwise declared to insurers ... and performed'. In the event of any dispute or disagreement over the interpretation, the president of RICS, or his or her nominee, will be the final arbiter.

The numerous references to asbestos in the various guidance notes and other RICS publications leaves no doubt that many of the services routinely provided by surveyors entail at least some degree of involvement with asbestos.

There is a specific exclusion for 'any claim or loss arising out of the death or bodily injury or disease of an employee under a contract of service with the firm(s) whilst in the course of employment or on behalf of the insured'. Similarly, there are exclusions for 'any claim or loss (including loss of value) arising directly or indirectly from pollution'.

This does not apply where such claim or loss arises from the insured's negligent structural design or specification, or failure to report a structural defect in a property; however, cover shall only extend to that part of any claim or loss that relates to the cost of redesigning, re-specifying, remedying and/or rectifying the defective structure, not including the cost of remedying and/or rectifying any loss or damage to the land.

It must be said that generally, to date, when handling claims, most insurers have chosen not to exclude liability for those claims arising (wholly or partially) from asbestos-related services. Nevertheless, the opportunity to do so is there, and dependent on the level of claims and the perceived risk, the attitude of insurers may well change in future.

It cannot be argued, under the *Unfair Contract Terms Act* 1977 (*UCTA*), that such exclusions, of what may be routine and common services, are unfair. This Act expressly excludes contracts of insurance, and in any case will not apply, because the insurance is taken out in a business or professional capacity, rather than as a 'consumer' (see section 6 'Limitations and exclusions'). Consequently, it is open to insurers to exclude liability for both personal injury and for economic loss arising from negligent errors and omissions by surveyors in their surveys, and the provision of advice or other services relating to asbestos.

More significantly, without professional indemnity insurance cover for all of the services they provide, surveyors will be guilty of a serious breach of RICS Rules of Conduct, with all of the implications that are attached to this. A number of surveyors may therefore be caught in a 'Catch 22' situation, where by virtue of historic practice and RICS guidance notes, they are expected to include asbestos as one of the numerous matters that they address in the course of their normal business, and yet they are specifically excluded from any professional indemnity insurance cover for the provision of these services.

The way forward?

One of the purposes of this guidance note is to clarify the small part that asbestos plays in the scope of services commonly provided by RICS members, as well as to set out the minimum standards of knowledge and performance that should reasonably be expected of each member. This note aims to distinguish these from the more detailed and specialist scope of services provided by 'asbestos inspectors' or advisors.

RICS is working together with the Asbestos Removal Contractors Association (ARCA) to develop a certification scheme for 'asbestos inspectors' that will be open to all, but will encompass the principle of RICS professional standards of practice and conduct, as well as a requirement for appropriate professional indemnity insurance.

In conjunction with this, RICS is talking to insurers and the government regarding the setting up of a ring-fenced insurance scheme for these specialist services. This would have a reasonable maximum limit on liability, to discourage unfounded or vexatious claims and the considerable costs involved in their defence.

Part 8
Asbestos inspection

8.1 Background

The employer, occupant or 'person in control of the premises' is required by law to be aware of significant hazards that may endanger their employees, visitors or the general public, and to take reasonable measures to prevent or control these (see Appendix 4 'Regulations'). This includes knowledge of the presence and condition of any 'asbestos-containing materials' (ACMs) in buildings for which they are responsible. It is this that gives rise to the need for an 'asbestos inspection'.

The *Control of Asbestos at Work Regulations* (*CAWR*) do not use the term 'survey', but instead refer to making an assessment as to whether asbestos is or is liable to be present in the premises. The Regulations do, however, require an inspection to be made of those parts of the premises that are reasonably accessible, with a consideration of the condition of any asbestos that is, or has been assumed to be, present.

It should be noted that these requirements are taken from Regulation 4 of the *CAWR*, which comes into force on 21 May 2004.

Historically, there has been no uniformly recognized procedure for conducting surveys for ACMs or for interpreting and reporting the findings. There is therefore often considerable variation in the level of service and form of documentation provided by 'asbestos inspectors'. In July 2001 the Health and Safety Executive (HSE) published guidance in the form of MDHS (100), entitled *Surveying, sampling and assessment of asbestos-containing material.*

The scope and nature of an inspection will vary according to the purpose for which it is required. The information may be needed in connection with specific activities, such as a proposed refurbishment or demolition, and may apply to the whole or only part of a premises, or, as is more likely, to the production of a plan to manage asbestos in non-domestic premises to comply with Regulation 4 of the *CAWR*.

Notwithstanding this, the aim of an 'asbestos inspection' is, as far as is reasonably practicable, to locate and report on *all* of the ACMs present in a building, so that the risks can be assessed and managed.

In new or recent buildings, the 'inspection' may merely comprise a desk-top review of the health and safety file, or other contract documents, for confirmation from the designers or builders that no ACMs were specified or used in the construction of the building. In the remainder and majority of buildings, however, a far more detailed investigation will be required.

The age and type of a building can give some initial pointers as to the likelihood of the presence of asbestos, and the locations of ACMs, but these cannot be relied upon. Even in properties of an early age, in which the use of asbestos was unlikely – for instance, Georgian houses – there is the possibility that later modifications may have incorporated ACMs, for example, as fire protection or thermal or sound insulation.

As described in Appendix 3, in the sub-section entitled 'Forms and products', ACMs have been used extensively for many different purposes and in a variety of common building products. Many of these may not be immediately obvious and are likely to be found in roof spaces, cellars, plant rooms, hidden in services ducts, under floors, above ceilings or behind decorative or protective finishes, panels and fittings. Unless the inspector is given unfettered access to look in every conceivable 'nook and cranny', to remove fixed panels, take up floor and ceiling coverings and generally take the building apart, he or she cannot guarantee to find every ACM.

In addition, although the chart in 'Forms and products' lists the majority of ACMs commonly used, it is not comprehensive. The full list is not known, and new and obscure uses for this amazingly versatile material continue to be discovered.

A comprehensive survey would often be disruptive and destructive, involving a degree of disturbance that may not be generally acceptable or reasonable. A compromise must often be found, and the client should accept that without virtual demolition, it may not be possible to find and identify every ACM. There is therefore a risk of subsequently discovering additional ACMs. That risk will be in inverse proportion to the extent of free access the inspector is permitted.

Sometimes the best, or only practical way, may be for an inspection to be carried out in several phases, with the inspector returning to the site to review parts of the building that are revealed during the construction or demolition works.

The enforcing authorities understand the dilemma, and if the client or inspector is prosecuted for endangering health due to a lack or inadequacy of information on ACMs, he or she will be judged on what was 'reasonably practicable' at that time.

8.2   Identification of ACMs

Although different types of asbestos are often referred to by their colour – white, blue or brown – this is not a reliable indicator. Sometimes, due to a shortage of supply or a glut of particular types, coloured dyes were used by unscrupulous suppliers and different types of asbestos mixed together.

By inspection alone, an experienced inspector, familiar with the full range of asbestos products, should be able to make an informed presumption that a material may contain asbestos, or alternatively, that it does not. However, the only way to be sure is by analysis of representative samples of the material, using a microscope.

In the absence of any analytical or other evidence to support a reasoned argument that they are highly unlikely to be ACMs, materials which have a similar appearance to them must be presumed to contain asbestos.

Unless a survey is supported by sampling (see section 9 'Sampling asbestos'), an inspector can only make 'presumptions' as to the presence or absence of ACMs. The presumption can either be 'strong', where there is good reason to believe that a material contains asbestos but this has not yet been confirmed by sampling, or as the 'default' position, when there is insufficient evidence to be sure that a material does not contain asbestos.

There are various valid reasons why it may not be appropriate or desirable for samples to be taken. In these cases, an inspector can only make unsubstantiated presumptions, based on his or her experience and the physical evidence to hand. The client must therefore accept the risk that not all of the asbestos may be identified, and also that suspect materials, which may eventually be proven to be asbestos-free, will need to be managed as though they are asbestos, with all the costs and disruption this may entail.

Under Regulation 4 of the *CAWR* there is a requirement to presume that a material contains asbestos and to manage it accordingly, unless there is reasonable evidence to the contrary.

8.3   Types of 'survey'

In the following section the term 'survey' is used as in the official document MDHS 100. However, in the context of this guidance note the term means 'asbestos inspection' and should be read as such.

MDHS 100 *Surveying, sampling and assessment of asbestos-containing materials* refers to three types of 'survey':

● Type 1: Location and assessment survey ('Presumptive survey');

● Type 2: Standard sampling, identification and assessment survey ('Sampling survey');

● Type 3: Full access sampling and identification survey ('Pre-demolition/major refurbishment survey').

The Type 1 survey

The purpose of a Type 1 survey is to locate, as far as reasonably practicable, any materials suspected of containing asbestos, and to assess the risk. However, this is limited to 'presumption' ('strong' or otherwise) and defers the need to sample and analyse until a later time.

All materials that are presumed to contain asbestos must be assessed.

The Type 2 survey

The purpose and procedures used are the same as for Type 1, except that this type includes representative sampling and analysis to confirm or refute the surveyor's judgement and presumptions.

The Type 3 survey

This is used in cases where major refurbishment or demolition is contemplated, and it is therefore reasonable and essential to provide full access to all areas, involving destructive inspection as necessary.

There is still the possibility that some asbestos may remain concealed, perhaps within chimneys or the structure, and will not be revealed until the works are under way. In such cases, the survey should be carried out in stages. In any event, emergency plans and arrangements should be in place to deal with any subsequent unexpected discovery of ACMs.

8.4 Briefing for an inspection

The following matters should be considered by anyone contemplating undertaking, or instructing another party to undertake, an inspection for asbestos.

Question/topic	Consideration
Purpose of the survey? (Type 1, 2 or 3 survey?)	For the general duty to manage, or for a more specific purpose, such as prior to refurbishment or demolition? This consideration will have an impact on the scope and content of the report
What existing information is available?	Health and safety file, previous asbestos reports or registers, drawings, specifications for a 'desk-top' survey
Limitations?	Inaccessible rooms/voids or other parts of the premises that are specifically excluded. (These must be identified in the report and/or plan)
Safety precautions?	For asbestos and for other hazards (for example, those created by the use of the premises, including by vehicles, processes, machinery and live services)
Sampling strategy?	Is sampling required? (See section 9 'Sampling asbestos')
Format and content for recording and presentation of data?	House style, specific requirements, drawings, photographs, labelling
Assessment and prioritization of risks of ACMs?	Categorization to be used, e.g., intact; damaged but not releasing fibres; or damaged and likely to release fibres
Management plan?	Who will prepare a plan for managing any risks arising, and review and update it? (The dutyholder must be involved)
Competence and Quality Assurance checks and procedures?	Accreditation, certification and peer review
Information to enable the assessment of the risks involved in conducting the survey?	First aid and welfare facilities, working arrangements, access equipment, site rules and procedures
Commercial matters?	Fees, expenses, costs of samples, additional unforeseen works/visits, programme for delivery of information

It is recommended that as part of the briefing, the inspector should attend a preliminary meeting to discuss the above and should 'walk through' the premises with a person familiar with the parts to be inspected, so that the difficulties involved in accessing or sampling may be highlighted and resolved.

All of the above, together with any other relevant details, should be recorded and sent to the client as confirmation of the instruction, and used as, or incorporated within, a written inspection plan.

8.5   Inspection data

The following information is required for each suspected ACM.

Topic	Consideration
Location	Building, room, floor level, position
Use of the room/area	Purpose and frequency of use
Asbestos type	White, blue or brown
Level of identification	Presumed (by default), strongly presumed, or proven by analysis of sample(s) or record documents
Extent	Area, length, volume, number, diameter
Asbestos content*	Approximate percentage content (often given as low, medium, high)
Product type	Wall panel, ceiling tile, pipe coating
Density*	If necessary, to distinguish between asbestos cement and asbestos insulating board
Accessibility	Concealed, exposed, internal, external
Surface treatment	Painted, sealed, unsealed, protected
Condition	Extent of damage or deterioration, using agreed categorization
Sample reference numbers	If sampled
Other relevant factors	Possible contamination of other parts, for example, part of a common air duct.

Ventilation, dispersal/dilution of airborne fibres, for example, externally in the open air |

*It may be necessary to give the asbestos content and density in order to make a judgement on the 'product type' and the regulations applicable.

The inspection report must be clear and unambiguous and should be in a form in which the whole or abstracts from it can be used as the basis of an asbestos register or log (see sub-section 10.4 'Asbestos registers or logs') or management plan (see section 10 'Management plan'). This will need to be kept readily available for on-site day-to-day use, in a form that can be updated and revised as necessary.

The management procedures of the premises should require that the register, log or management plan is always referred to prior to the authorization of any maintenance or building works. Ideally, this should form part of a formal 'permit to work' system.

8.6 Extract from '10 Key Facts' in *A Comprehensive Guide to Managing Asbestos in Premises*, HSE Books

- 'Regulation 4 of the *CAWR* is a duty to manage, not a duty to survey.'

- 'You don't always have to do a survey, but you do have to manage your ACMs.'

- 'A bad survey is worse than not carrying out a survey.'

8.7 Lessons learnt from experience

Asbestos sheeting generally contains chrysotile (white asbestos). However, in times of short supply, crocidolite (blue asbestos) was sometimes used.

The presence of (reflective) mica flakes is not necessarily an indication of recent asbestos-free mineral fibre boards or tiles, as these flakes were also used in some earlier compositions that do contain asbestos.

Asbestos-stripping techniques and monitoring procedures have improved with time. Sometimes, although an area may appear to have been stripped in the past, poor workmanship may have resulted in asbestos debris being left behind on pipes, structural steelwork and plant, or concealed beneath asbestos-free insulation that has subsequently been installed.

Part 9
Sampling asbestos

9.1 General

If carrying out a Type 2 or 3 'survey' (see section 8 'Asbestos inspection'), in accordance with the Health and Safety Executive (HSE) guidance note MDHS 100, then sampling is an essential and integral part of the 'survey'.

Types of sampling

There are commonly two types of sampling used in asbestos: bulk sampling and air sampling.

The former involves taking physical samples of materials that are suspected of containing asbestos (for example, ceiling tiles, wall panels and pipe lagging). The latter involves the use of specialist pumps to filter a known volume of air and thus determine the number of airborne respirable fibres.

In both cases, the samples require analysis by skilled laboratories, using optical microscopy. In special circumstances, for example, to distinguish conclusively between different types of fibre, electron microscopy may be needed.

9.2 Bulk sampling

In bulk sampling, the skilled laboratory technician determines the type and an approximation of the percentage content of asbestos by microscopic examination of the unique physical characteristics of the sample.

The United Kingdom Accreditation Service (UKAS) accredits laboratories for bulk sampling, fibre identification and fibre counting, but not for determining the approximate proportion of asbestos content. In addition, there is currently no uniform common terminology used among laboratories for reporting their results.

Regulation 20 of the *Control of Asbestos at Work Regulations (CAWR)*, which comes into force in November 2004, states that only persons accredited as complying with ISO 17025 may be employed to analyse a sample of material to determine whether it contains asbestos.

The physical operation of taking bulk samples falls within the *CAWR*. The operation is subject to stringent precautions in view of the potential health risks that can arise both for the person taking the sample and for the occupants of a building or the general public, who otherwise may be affected by contamination during the process.

Bulk sampling strategy

The sampling strategy must be discussed and agreed with the client beforehand.

Question/topic	Consideration
Purpose of samples	As part of a Type 2 or 3 asbestos survey; for a management plan; or limited to a particular area or material
Limitations/restrictions	Areas that are not to be disturbed, such as dealer rooms or the board room
Competence	Accreditation and relevant experience
Working arrangements	Out of normal working hours?
Known hazards	For example, fragile roofs, or other hazardous substances or processes
Number of samples	Representative of suspected asbestos-containing materials (ACMs)
Type of materials to be sampled	Coating, board, loose
Information required from samples	Type and approximate percentage or proportion of asbestos Density of board material?*
Method of taking samples	Size of samples
Access arrangements/facilities	Stepladder/use of hydraulic platforms
Possible impact on business	Parts may need to be temporarily closed down
Protective measures	Enclosure, shadow-vacuuming, dust sheets, cleaning-up of debris
Exclusion of unprotected personnel	Warning notices, ropes or barrier tape
Emergency measures	In case of an incident causing extensive contamination
Repairs	Both temporary and permanent?
Labelling of areas sampled	Labels may be impracticable, undesirable or cause unnecessary alarm. Unobtrusive colour coding may be an alternative
Contingency	Possible need for return visit if results are not conclusive
Reporting	Programme and presentation of results

*If the material being sampled is a board, then it may be necessary to establish the density of the material, to determine whether it is 'asbestos cement' or 'asbestos insulating board'. Different regulations apply to each.

Care is required to ensure that there is no cross-contamination of the sample(s). Where a number of samples have been taken, there should be a comprehensive referencing system, to ensure that the results refer to the correct locations and that the samples are properly stored for a suitable period, so that they can be double-checked by others at a later date, if necessary.

In all, there must be confidence in the integrity of the service provider and its systems, which can only be achieved by the use of an appropriately accredited laboratory.

How many samples?

The principle is to obtain representative samples of each ACM in sufficient number to take into account any apparent variations in the suspect material, including changes in its depth.

Homogenous manufactured products or components such as boards, sheets, cement, pipes, textiles, ropes, plastics and vinyls, bitumen roofing felt and gaskets may be reasonably assumed to be similar, with the asbestos content uniformly distributed. One or two samples of these may suffice. Insulation and spray coatings that were applied on site, however, are not always consistent. The type and percentage of asbestos present in these often depends upon what material was available at the time.

Other factors to be taken into account include the likelihood of previous repairs or replacements; the possibility of different ages of components (for example, ceiling tiles, some of which may have come from old stock-piles); and the amount of the suspect materials or the number of components (insulating boards or tiles, for instance, may require one sample to be taken per room or every 25m^2).

The client's planned and subsequent activities may also be relevant. For instance, where the whole or part of a building is due for refurbishment or demolition, it may be necessary to take samples to 'prove' conclusively that materials or components do not contain asbestos, as well as to establish those that do.

The selection of samples is subjective, but should be made on the basis of a close visual inspection, looking for changes in colour, thickness, surface finish, repair and tell-tale identification marks or trade names.*

Records

As it is taken, each sample must be double-bagged before being passed to the laboratory for analysis. The sample should be labelled with a unique identification, which is also recorded in the inspection documentation, in order for its origin to be traceable. As much information should be recorded as possible, to ensure that there is no confusion as to the source location (in case labels fall off or are removed).

Risk assessment

Prior to sampling, a risk assessment should be conducted, taking into account the agreed sampling strategy and any significant hazards identified during the briefing with the client. From this, a safe system of working should be developed and recorded, via a written method statement or plan of work.

*See Appendix 3 for a note on trade names.

The following issues need to be addressed:

a) protection of the sampler

- **PPE:** an assessment should be made of the personal protective equipment (PPE) required – for example, disposable overalls, overshoes and a suitable mask or respirator (dictated by the type of material and the extent of disturbance, or the area from which the sample is taken, such as shafts or ducts).

- **Dust suppression:** the PPE is a last resort. Airborne emissions should be controlled and minimized by pre-wetting the sample with water or a suitable wetting agent. Boards or sheets may be surface-sprayed, but lagging may need injecting. If this is not likely to be totally effective, or if it is not safe to use liquid (perhaps because of proximity to electricity), then 'shadow vacuuming', using a type H (BS 5415) vacuum cleaner, should be carried out.

b) protection of the occupants

The following points need to be taken into consideration:

- vacation of occupants from areas to be sampled;

- protection of surfaces and equipment with impervious, easy-clean cover sheets;

- removal and disposal of contaminated debris;

- sealing or making good sampling points, to prevent further fibre release (with PVA spray, tapes and/or plaster);

- prevention of inadvertent disturbance of suspected ACMs until proven to be safe (via warnings, management procedure, physical protection or barriers).

9.3 Air sampling

Regulation 19 of the *CAWR* states that only those accredited to ISO 17025 or equivalent criteria can conduct air testing. The process involves the collection of fibres using a filter, which is then microscopically examined to count the number of fibres within a specified number of fields (with sub-divisions marked on a slide), to assess the likely number of respirable asbestos fibres.

Why air sample?

There are two main reasons why air sampling is required.

The first is in connection with the process of removing, treating or carrying out other major work that will disturb the asbestos. Concurrent air sampling may be used to determine exposure of the workers to asbestos fibres, in order to assess the adequacy of control measures. It may also be used outside of working enclosures to check whether any airborne fibres have escaped, and thus monitor the integrity of the airtight seals.

The second reason is for 'reassurance testing', where air sampling is used to determine whether any airborne fibres are present prior to removal of the enclosure and reoccupation of an area that has been subject to asbestos stripping or other form of decontamination.

The clearance reading that must be achieved is 0.01 fibres per ml of air. It should be appreciated that this has been accepted to be the reasonable limit of the optical microscope typically used by laboratories. It is a pragmatic assessment, not a solemn declaration of safety, as there is no recognized safe limit for airborne asbestos fibres.

The reassurance testing is the last operation in a sequence of checking, which includes a close visual inspection to ensure that no dust or debris remains. The procedures described in the Health and Safety Executive (HSE) guidance note MDHS 39/4, which sets out good practice, should be followed to ensure that the results are accurate and realistic. For example, the air pumps should be warmed up and be running efficiently prior to sampling; filter holders should ideally be fixed face-downwards 1–2 metres above the floor; and the typical working air movement or disturbance should be recreated, so that the testing conditions are realistic.

Often, the existing environment is already 'naturally' contaminated, perhaps by asbestos fibres within the general external atmosphere (possibly from vehicular brake linings), or by the presence of other fibres or dust sources, such as carpets or fibreglass. This may result in artificially high readings. It is therefore useful, and sometimes essential, for 'background readings' to be taken, to establish a point of reference prior to the actual sampling process.

The analysis of air samples is conducted following a rigid and uniform regime, the procedures of which insist that any fibre falling within a stated size category (comprising diameter and length and the ratio of both) must be deemed to be asbestos and counted as such. The final figure is derived by a statistical calculation of a representative selection of fields on the slide on which the sample is mounted.

Other purposes

There may be situations where it is necessary to distinguish between actual, rather than nominal, asbestos fibres, and other similar but less harmful fibres, such as man-made mineral fibres. In these cases the normal counting procedures are inappropriate and may be misleading.

In such cases, special arrangements need to be agreed outside of the accreditation scheme. It may also be necessary to use more powerful optical equipment, such as an electron microscope, the costs of which generally restrict their availability to universities or similar institutions.

9.4 UKAS guidance

UKAS has issued the following guidance in connection with sampling:

- air sampling for 'clearance indicator' testing should only be undertaken in dry conditions and should not be carried out where stripped surfaces have been sealed;

- a visual inspection of the area to be sampled, to check for remaining asbestos debris, is necessary before air sampling;

- cross-contamination must be avoided by ensuring that there is sufficient equipment, or a means of cleaning equipment between uses;

- samples taken by a laboratory should usually be analysed by that laboratory. In any case, they should always be analysed by a laboratory holding UKAS accreditation for that particular test;

- there is no UK standard test specification for identification. The HSE guidance note MDHS 77 describes the characteristics of asbestos and may be used as the basis of an accredited method;

- laboratories should have documented in-house procedures and training, to enable identification of the six regulated types of asbestos. The six types are crocidolite, amosite, chrysotile, fibrous actinolite, fibrous anthophyllite and fibrous tremolite. Identification can be achieved by holding known reference samples of each for comparison. The latter three types are only rarely found in buildings;

- the use of photographic records is strongly recommended, in order to demonstrate the condition and location of suspected ACMs at the time of sampling or surveying;

- the counting of fibre samples and identifying of asbestos components is subjective and liable to human error, relying upon operator experience, training and procedures. In order to minimize uncertainty, comprehensive records must be retained and regular checking and auditing procedures undertaken, to maintain quality control.

MDHS 39/4 requires laboratories that carry out fibre-counting to participate in the Regular Interlaboratory Counting Exchanges (RICE) scheme, which is run under the auspices of the Committee on Fibre Measurement (CFM). In this scheme, sample slides are periodically re-examined by other laboratories.

Although there is not yet a similar compulsion for the identification of asbestos in bulk materials, UKAS states that laboratories should participate in appropriate inter-laboratory comparison or proficiency testing programmes, such as the Asbestos in Materials (AIMS) scheme, which is also run under the auspices of the CFM.

9.5 Qualifications

The British Occupational Hygiene Society (BOHS), in conjunction with the HSE and industry representatives, has developed modules that have relevance for asbestos sampling. For details of these, see Appendix 7.

Part 10
Management plan

10.1 General

Regulation 4 of the *Control of Asbestos at Work Regulations* (*CAWR*) requires every 'dutyholder' of non-domestic premises to assess whether asbestos is, or is liable to be, present; to prepare and implement a plan for managing any risks arising; and to review and revise the plan as necessary. This duty does not come into force until 21 May 2004. (See Appendix 4 'Regulations' for a summary of the requirements and an explanation of the terms used.)

The Health and Safety Executive (HSE) has published *A Comprehensive Guide to Managing Asbestos in Premises*. This lists the following 'Key Facts' in respect of the plan:

- 'Managing asbestos means maintaining your asbestos containing materials (ACMs) in good condition to protect two groups of people :

  - those who work on the fabric of the building (electricians, plumbers etc.); and

  - those who work in the building (e.g. plant and office workers, cleaners etc.)

  and may come into contact or work near damaged or deteriorated ACMs';

- the plan 'is your way of ensuring that your employees or others do not disturb your ACMs';

- the plan 'can take many forms and need not be complex, but it does need to be effective'.

The management plan is an important legal document. In addition to its health and safety significance, it is required to be made available to, and to be inspected by, a variety of interested parties, including:

- those who intend to inspect or carry out works to the premises, who will need to know the location and condition of asbestos in order to safeguard their own health and safety;

- valuers, surveyors and those who prepare schedules of condition and dilapidations. Depending on the scope of their commissions, they will need to review the contents of the plan, to be able properly to advise their clients on the presence of ACMs and their likely financial and other implications;

- solicitors acting for potential purchasers or tenants. The plan will be added to the list of essential documents that they will wish to obtain as part of their routine legal searches;

- occupants and employees' representatives, to enable them to be aware of any risks to their health;

- contractors, caretakers, facilities managers and managing agents carrying out or organizing building alterations or maintenance;

- the emergency services (fire brigade);

- landlords and head lessees, to ensure that their tenants are adhering to lease covenants requiring compliance with statutory requirements; and

- HSE inspectors checking on compliance with Regulation 4 of the *CAWR*.

The absence of a suitable plan may thus have significant financial implications or affect the liquidity of the premises as an asset.

A plan is not required when the assessment confirms that asbestos is not or is unlikely to be present in the premises – for example, in a building constructed after 1999, with confirmation from the project team that asbestos has not been used in its construction. Nevertheless, a record should be kept of the assessment carried out, and its conclusion, to show to an HSE inspector or a prospective purchaser or occupant.

10.2 Content

This guidance note is not intended to dictate or specify precisely how such a plan should be prepared, or its format and content. For advice on this, readers should refer to the HSE publication *A Comprehensive Guide to Managing Asbestos in Premises*. It is, however, important that RICS members and their clients understand the basic components of the plan and the factors to be considered in its preparation, so as to be able to advise, comment or brief others as necessary. The principal matters are set out in the table below, which indicates questions to be asked and considerations to be taken into account when preparing or reviewing a plan.

Question/topic	Consideration
Competence of author	Should be accredited or certified, or otherwise suitably experienced and equipped
Dates of plan	When first prepared and last reviewed
Working limitations	Access restrictions caused by operations of occupants, plant and machinery, or because the materials are concealed by shopfitting, linings or within ducts, etc.
Extent of premises inspected and included	All parts reasonably accessible, including outbuildings, fixed and mobile plant, external pipes and bridges, service risers and ducts, above ceilings and below floors
Is presence or absence of asbestos 'proven' or 'presumed' (or 'strongly' presumed)?	If 'proven', then by what means (for example, shown on as-built drawings, in the health and safety file, or proven through sampling). If 'presumed', then what factors were considered (for example, physical appearance, typical use and age)
Sampling	Competence of sampler, sampling regime and scope (from 21 November 2004 samples may only be analysed by an accredited person or organization)
Dutyholders' intentions for premises	Take account of dutyholders' short, medium and long-term plans, such as redecoration, refurbishment or demolition

Significant changes	The plan must take account of all significant changes to the premises since the initial preparation or last review of the plan
Measures for managing risk	Measures to protect, maintain or, where necessary, remove asbestos, must be specified, including information on the location and condition of ACMs
Means of providing information on location and condition to persons likely to disturb asbestos	Labelling of ACMs, or 'permit-to-work', requiring inspection of the plan prior to any works that could disturb asbestos (such as maintenance, alteration or demolition)
Distribution of significant information	To occupants and safety representatives of employees
Emergency provisions and liaison with emergency services	Emergency arrangements; information to be made available to the fire brigade
Availability of the plan	Details of location and person responsible for its safekeeping. Plan should be kept on site, unless premises are vacant. In this case, the plan should be kept nearby
Monitoring (condition of ACMs)	Arrangements for monitoring ACMs – by whom and how often? (frequency depends on individual circumstances and the likelihood of damage, but should be checked as a minimum every six to 12 months)
Reviewing (implementation of plan)	Arrangements for reviewing management measures – periodic checks should be made to ensure that arrangements are working, for example, when new staff are appointed, and, as a minimum, every six months. The extent of the review will vary according to the type, size and complexity of the premises
Records	Conclusions of each monitoring and review should be recorded

The dutyholder owns and is responsible for the safekeeping of the plan. However, he or she is obliged to make the information available to anyone who is likely to disturb asbestos, and this includes new owners or occupants. The plan must be kept available for the life of the premises.

The dutyholder must be involved and take responsibility for the preparation of the plan, as only he or she will be aware of his or her future intentions and anticipated programme for redecoration, refurbishment or major alterations. Such proposals, which could disturb ACMs, will need to be recorded and taken into account in the plan.

10.3 Ways of managing ACMs

The risks arising from ACMs can be controlled or removed, either indirectly by management systems or 'management actions', or directly by physical activities or 'control actions', or by a combination of both.

(INDIRECT) – 'MANAGEMENT ACTION'

Method	Detail
Isolate and restrict or exclude access	Restricted zones with controlled entry, behind locked doors with keys available only to authorized persons
Warn of dangers	
● Signs or labels	Signs or labels must be clear and durable (and should be a last resort, if no other appropriate preventative or protective measures can be taken). Labelling should conform with the *Health and Safety (Safety Signs and Signals) Regulations 1996*
● Training of staff	Training of staff could form part of inductions, reinforced by periodic 'toolbox talks'
Monitor condition	Periodic inspection (as a minimum, every six–12 months), with condition recorded
Permit to work	Formal procedure to ensure that prior to any activity that might disturb asbestos, reference is made to the asbestos register or (after 21 May 2004), the management plan. This should be included within the 'site rules' issued to contractors
Maintain and update record of ACMs	Formal procedure to ensure that any alterations are recorded and that the asbestos register or (after 21 May 2004) the management plan is updated

(DIRECT) – PHYSICAL 'CONTROL ACTION'

Method	Detail
Decontaminate	Only by competent persons, using appropriate control measures and suitable equipment
Repair damaged ACMs	Use appropriate fillers and coatings
Enclose/protect	Use physical barriers or cover panels (take care when fixing to ACMs)
Encapsulate:	To prevent release of airborne fibres:
● paint	Asbestos cement – use alkali-resistant paint Insulation board – use PVA emulsion paint
● proprietory	Bituminous or flexible/semi-flexible polymeric coating
Remove	Only by competent contractors*, using appropriate control measures and equipment

*(Must be licensed by the HSE unless undertaking only 'minor works' or works involving asbestos cement, floor tiles or bituminous materials)

The publication *Asbestos and Man-made Mineral Fibres in Buildings*, produced by the Department of the Environment, Transport and the Regions in August 1999, contains useful advice, with flow charts, to assist in the selection of appropriate control actions. This is widely accepted as the current official guidance.

10.4 Asbestos registers or logs

Non-domestic premises

Prior to the *CAWR* 2002, there was no specific legal requirement to produce a register, log or record of asbestos materials in buildings.

There is, however, general health and safety legislation, notably the *Health and Safety at Work etc. Act* 1974 and more specifically, the *Management of Health and Safety at Work Regulations* 1999, which require 'employers' (including persons in control of premises) to assess all significant risks, including asbestos, to the health and safety of their employees and to anyone else who may be affected by their undertakings, and to remove or control these risks and record significant findings.

Regulation 4 of the *CAWR*, which comes into force on 21 May 2004, requires every dutyholder for non-domestic premises to assess whether asbestos is or is liable to be present, and where appropriate, to prepare and implement a plan for managing any risks arising, and to review and revise the plan as necessary.

The Regulation itself does not specifically refer to a 'register', but the guidance, in paragraph 29 in the context of example 3, states that 'a record of the asbestos present, such as a *register*, must be compiled and kept up to date...' This new requirement will inevitably result in the need for a record of asbestos to be kept for all non-domestic premises.

Domestic premises

Domestic premises are excluded from the *CAWR*, but all premises, including domestic premises, are subject to the *Occupiers Liability Acts* (1957 and 1984). These require property owners and occupiers to be aware of potential significant hazards to health and safety and to take such care as is reasonable in all circumstances, to ensure that all persons, including visitors and trespassers, do not suffer injury on the premises. The location of any asbestos, and its condition, needs to be taken into consideration in deciding what protective measures, if any, are required.

10.5 Emergency plan and procedures

Where asbestos is present, employers and 'persons in control' of premises are obliged to have an emergency plan prepared in advance, including the following.

Scope/subject	Requirement
Safety procedures	Ensure that safety procedures, including safety drills, have been prepared
Information	Ensure that details of the emergency arrangements, including hazard identification and a note of relevant work hazards and specific hazards likely to arise at the time, are available
Warning and communication systems	Establish warning and communication systems to facilitate an appropriate response in the event of an emergency
Inform	Ensure that details of the emergency arrangements and the warning and communication systems are made available to the relevant accident and emergency services (internal or external) and displayed at the workplace, if appropriate
Immediate response	Note the immediate steps to be taken to mitigate the effects, restore the situation to normal and inform any person who may be affected
Limit access	Restriction of access to affected areas to those persons needed to carry out remedial measures and who are properly trained, equipped and protected

This obligation also applies to vacant premises.

Part 11
Types of property

11.1 General

Asbestos can be found in almost all types of building: commercial, industrial, residential, educational, recreational and agricultural.

The age of a building may be an initial guide as to the likelihood of asbestos content, but it is not conclusive and cannot be relied upon. For example, asbestos would not usually have formed part of the original construction of historic buildings, but it may have been used in subsequent alterations, especially to services. It was particularly popular in the 1960s. Except for specialized filters and the like, its use has been illegal since 1999.

The type of construction of buildings can also give a clue. Steel-framed buildings, for instance, often used asbestos, either sprayed or in board form, as fire protection, whereas this was generally not necessary in those with concrete frames. In lightweight timber-framed buildings, particularly those that were system-built, asbestos boarding was often used for external cladding, as internal linings and sometimes as a fire-resisting core as well.

11.2 Residential

The majority of the legislation concerning asbestos arises from the parent enabling act, the *Health and Safety at Work etc. Act* 1974, and is thus confined to workplaces (that is, 'non-domestic premises').

The Act, and the regulations that flow from it, do not apply to 'domestic premises', namely 'a private dwelling in which a person lives', but do apply to any work activity that takes place there, for example, work undertaken by a builder, plumber, electrician, and so on.

There is more general legislation that applies to non-domestic premises, such as the *Defective Premises Act* 1972 in England and Wales, the *Government (Scotland) Act* 1982 in Scotland and the *Occupiers Liability Acts* (1957 and 1984).

In addition, legal precedents have established that the common parts of flats (in housing developments, blocks of flats and some residential conversions) are *not* part of a private dwelling, and are therefore classified as non-domestic.

Thus, Regulation 4 of the *Control of Asbestos at Work Regulations* (*CAWR*) 2002, which imposes a 'duty to manage asbestos' in all non-domestic premises, does not apply to houses or private dwellings, but does apply to any common parts of those premises.

Because of the wide variety of different forms of internal layouts, tenures, mixed uses and degree of permanence of these arrangements, there are bound to be grey areas. The final arbiter as to whether a particular premises or a part thereof is domestic or not will be the courts. In the meantime, RICS members will have to use their own judgement, or perhaps, in specific situations, seek guidance from their local Health and Safety Executive (HSE) office.

In order to give some assistance, a list of typical residential configurations and an indication of whether these are probably domestic or non-domestic is shown in chart form in Appendix 8. This is based on information contained in the Approved Code of Practice (ACOP) to Regulation 4 of the *CAWR* 2002 and the HSE's response to questions put by RICS. This list is non-exhaustive, but should give an indication of the principles to be considered in most cases.

It can be seen from the above that the legal duties in respect of asbestos vary according to the particular configuration or use of the premises, and this must be taken into account. In many cases, the legal situation will be obvious, but in others it may not. An assumption will need to be made.

The surveyor may not be aware of the client's or occupant's intended short- or long-term use of the premises. It is therefore important that not only are any assumptions and stated intentions recorded in any report, but also, where appropriate, that a warning is included that any subsequent change of use may alter the legal responsibility in respect of asbestos. Further professional advice should be sought at this time.

11.3 Vacant premises

If the premises are vacant, the dutyholder, whether this is the landlord or tenant or a combination of the two, still has a duty to manage asbestos to the same extent as if the premises were occupied.

People likely to be at risk in vacant premises include security staff, cleaners, agents and potential purchasers or lessees visiting, surveyors inspecting on behalf of others, trespassers and the emergency services.

There is no requirement to ensure that asbestos management plans (see section 10 'Management plan') are kept on site, but they must be kept within a reasonable proximity and be readily available.

Part 12
Asbestos removal

12.1 Basic principles

The basic principles with regard to the safe removal of asbestos are the following:

- use of trained personnel;

- strictly controlled conditions – to safeguard occupants and the general public, as well as workers;

- protection of personnel (monitors as well as asbestos strippers);

- safe disposal of contaminated waste;

- maintenance of comprehensive records;

- time implications; and

- other sundry factors.

These are discussed below.

Trained personnel

- Only competent, adequately resourced personnel (licensed contractors and accredited laboratories – see Appendix 7) should be used.

Strictly controlled conditions

- The working area should be enclosed. Use dust-tight, suitable impervious sheeting, supported by a framework, fixed or adhered to the wall, ceiling or joinery, with all openings and air vents sealed and three-stage air locks for entry and exit.

  Incorporate clear vision panels for monitoring purposes.

  (Enclosures may not be practical for use in external situations, or special weather protection may be required.)

- Equipment, fittings, floor coverings and so on should be protected, to prevent contamination.

- Air extraction – where reasonably practicable – should be carried out, to generate negative air pressure, reduce dust levels and prevent dust escape through any imperfections in the enclosure or air locks. Use high-efficiency filters, with the exhaust vented to the external air.

  Extraction should be calculated according to the air volume of the enclosure.

  Overrun fans as long as possible, to remove airborne fibres.

- Conduct the following monitoring:

  - background air monitoring – to provide base datum;

  - visual inspection/smoke test of enclosure for leaks, prior to the start of stripping;

  - leakage air monitoring during stripping (one-two hours after start of stripping);

  - on completion, visual inspections for any remaining dust or debris (paying particular attention to ledges, recesses and voids);

  - clearance air monitoring within the enclosure, to see if safe to remove;

  - possibly reassurance air monitoring outside of the enclosure, to see if safe to reoccupy; and

  - thorough cleaning of surfaces after removal.

- Clean within the enclosure, taking note of the following:

  - surfaces should be wire-brushed, where appropriate;

  - use vacuum cleaners with high-efficiency filters;

  - use tack rags to wipe down all surfaces, ledges and so on;

  - all surfaces (including the enclosure) should be sealed with PVA, or similar;

  - all tools, dirty overalls, respiratory equipment and so on, should be cleaned before removal, then sealed in impervious bags, prior to exit via air locks.

Protection of personnel

- Personnel should wear impervious protective clothing – close-fitting at cuffs and ankles, with no pockets – as well as head covering, easy-clean footwear or overshoes. If there is a need to wear wellington boots, overalls should be dressed over rather than tucked into them.

- Use differently coloured sets of overalls for the various stages of decontamination and transit to remote decontamination facilities. This will readily demonstrate that the proper procedures are being followed.

- Use respirators with appropriate filters, or breathing apparatus (personalized to suit the wearer's physical characteristics – size and shape of face, facial hair, etc. – in order to ensure an airtight seal).

- Hygiene facilities must be provided, including accommodation for decontamination and showers (mobile, or purpose-built on site). Power, water and filtered drainage should be available.

Safe disposal of contaminated waste

- Consult with the disposal authority – in England, the county council; in Scotland or Wales, the district council; or the local Environment Agency office.

- Note that fibrous asbestos is classed as 'special waste' under the *Control of Pollution (Special Waste) Regulations* 1980.

- Remove material intact where possible. Remove or cut fixings.

- Double-seal material in two suitable plastic bags, or double-wrap in polythene. Each bag must be clearly and permanently marked with a suitable warning of contents, together with the name and address of the removal operator.

- Waste should be immediately removed from the site or stored in special sealed skips and kept locked.

- Maintain a strict system of records, with a 'consignment note' system for tracking removal from site and disposal at a licensed site.

- The usual method of disposal is burying or earth cover. Other recent patent systems include 'vitrification', where asbestos is 'melted' into glass.

Comprehensive records

Keep the following records:

- air-monitoring results;

- maintenance records of plant, change of filters, etc.;

- a note of any defects, leaks and so on, and action taken;

- copy of insurance policies (professional indemnity insurance and employer's liability);

- copy of the training records of strippers;

- copies of the medical certificates of strippers;

Retain sample slides for cross-checking if required.

Time implications

Factors to be considered include the following:

- out-of-hours working is common;

- flexibility – it is often necessary to stay longer than anticipated, to deal with unforeseen problems;

- preparation time is required for:

  - background air sample (taken while enclosure is erected); and

  - erection/sealing/testing of enclosures and air locks.

- stripping of asbestos-containing materials (ACMs). Consider the following:

  - wet-stripping (injection process);

  - cleaning of all surfaces (structure, fabric, finishes and enclosure);

  - limits on working hours, due to the high temperatures generated within protective suits and the lack of ventilation, and frequently due to the heat generated by the working plant and equipment;

  - limits to the number of sample readings per analyst; and

  - putting on of personal protective equipment (PPE) and time spent in the decontamination process.

- clearance and dismantling. Consider:
    - allowance for percentage failure of air tests and retesting for clearance and reassurance air sampling;
    - recleaning of interior of enclosure after each failed test; and
    - time for clearing away and making good.

Other sundry factors

The following must be taken into account:

- provision of temporary facilities/protection;
- informing the statutory authorities – fire brigade, police, etc.;
- erection of warning notices – the users of the building must be informed (use multilingual notices if appropriate);
- means of escape in case of fire;
- fire precautions – have portable fire extinguishers on hand;
- the importance of close liaison and good, quick communications between the analyst and strippers, to avoid delay in obtaining test results;
- provision of an on-site laboratory for the rapid analysis of air samples;
- use of a patent demountable framework for the enclosure, airlocks and so on;
- the re-use of enclosures – these will need thorough cleaning;
- use of patent enclosures, for example, incorporating gloves for the removal of small panels or limited sections of lagged pipework, etc.; and
- emergency procedures. These must be thought out and catered for in advance. Consider the level of contamination at which emergency action must be taken (including the vacation of occupants, provision of temporary accommodation, and so on).

12.2 Methods of removal

The following should be noted with regard to methods of removal.

- Dry stripping is no longer permitted. Use wet stripping (dampening). Consider use of wetting agents or surfactants.
- Use a multiple-point injection process.
- High-pressure water jets can be used.
- Ideally, materials should be dampened down to reduce fibre emission. However, electrical equipment must be protected. Waste water must be collected or filtered and the area must be allowed to dry out.
- Board material should be removed intact where possible, causing minimal disturbance, for example, by unscrewing, cropping or fixing bolts.
- Where possible, sections of plant or plant lagged with asbestos can be wrapped, cut and removed intact.

The HSE publication *Introduction to asbestos essentials* contains a table that shows typical exposures to asbestos fibres in cases where poor control measures and work practices have been employed. The exposures, measured in fibres per ml of air, range from 'up to 1,000' for dry removal of sprayed (limpet) coating to just '1' for hand-sawing or drilling of asbestos cement.

Hand-sawing of asbestos insulating board is likely to generate 10 times more fibres than similar works carried out to asbestos cement. The use of a powered saw increases the fibre release by a factor of between 2 and 20, depending on the type of saw. Meanwhile, just sweeping up debris from asbestos insulating board could generate up to 100 fibres per ml of air.

Appendix 1
Definitions and abbreviations used

Definitions

Action level | The accumulative exposure to asbestos over a stated period of time, measured in fibre-hours per ml of air, for particular forms of asbestos, which if exceeded triggers notification of the Health and Safety Executive (HSE), the designation of 'asbestos areas' and regular medical surveillance of employees.

Asbestos | The generic term for a group of naturally occurring fibrous mineral silicates. Asbestos is obtained by mining; the main producers are Canada, South Africa and parts of the former USSR.

In the *Control of Asbestos at Work Regulations* (*CAWR*) the term 'asbestos' includes crocidolite, amosite, chrysotile, fibrous tremolite or any mixture of those materials. The three significant types of asbestos that have been commercially used in the UK are:

CROCIDOLITE, commonly known as **BLUE**
AMOSITE, commonly known as **BROWN**
CHRYSOTILE, commonly known as **WHITE**

Asbestos area | Designated area, triggered by exceeding 'action levels', to which access is restricted to properly trained and equipped personnel, for the purposes of repairing, removing or treating asbestos.

Asbestos inspection | An inspection of buildings, structures, plant and land, conducted by an 'asbestos inspector', where the prime objective is to determine or assume the location, type and condition of materials containing asbestos.

Asbestos inspector | An asbestos inspector is a person or organization who conducts an 'asbestos inspection' of buildings, structures, plant or land, with the specific and single objective of identifying and reporting on asbestos-containing materials (ACMs). This infers the need for a particular qualification, which may become a statutory requirement in the future.

Client | Generally, this is used to describe the employer or person commissioning the inspection or survey. Where used in the context of the *Construction (Design and Management) Regulations* (*CDM*) 1994, it has a specific legal definition.

Contaminant	A substance (or substances) present above normal background levels, with the potential to give rise to adverse effects.
Contamination	In the RICS *Appraisal and Valuation Manual* (the *Red Book*), 'contamination' refers generally to concerns in the ground, and is associated with land contamination.
Control limit	The accumulative exposure to asbestos over one of two alternative stated periods of time, measured in fibre-hours per ml of air, for particular forms of asbestos, which if exceeded triggers the designation of 'respirator zones' and the mandatory use and maintenance of respiratory equipment. (See discussion of the *CAWR* in Appendix 4 'Regulations'.)
Deleterious materials	A material or component that has the capacity to cause harm.
	In the property world, this term is usually associated with building materials or components that can result in the deterioration of the building fabric or structure, or which constitute a risk to the health and safety of the occupants.
	Such materials are a financial risk to an investment in property; consequently, their use is often specifically prohibited by Funds and other experienced investors.
Dutyholder	The person or persons legally responsible for ensuring that a regulation is complied with. (See discussion of Regulation 4 of the *CAWR* 2002 in Appendix 4 'Regulations'.)
Hazard	Something with the potential to cause harm.
Respirator zone	Designated area, triggered by exceeding 'control limits', to which access is permitted only to personnel wearing appropriate respirators. (See discussion of the *CAWR* in Appendix 4 'Regulations'.)
Risk	The likelihood of harm occurring and the severity of the consequences.

Abbreviations

ACAD   Asbestos Control and Abatement Division of TICA

ACM   Asbestos-containing material

ACOP   Approved Code of Practice

AIC   Asbestos Information Centre Ltd

ARCA   Asbestos Removal Contractors Association

ATAC   Asbestos Testing and Consultancy wing of ARCA

BOHS   British Occupational Hygiene Society

CAWR   *Control of Asbestos at Work Regulations*

HSE   Health and Safety Executive (responsible for enforcing regulations). The Health and Safety Commission (HSC) is responsible for instigating regulations

HSV   RICS Homebuyer Survey & Valuation service

MDHS   Method of Determining Hazardous Substances

RICS   Royal Institution of Chartered Surveyors

TICA   Thermal Insulation Contractors Association

UKAS   United Kingdom Accreditation Service

Appendix 2

Health issues and risks arising from asbestos

Health implications

The physical characteristics of asbestos that render it hazardous to health are its sharp, crystalline, fibrous nature; its ability to split into a family of fine fibres; and the durability of these fibres.

The fibres are microscopic and are thus readily carried by air currents. They can remain airborne for a considerable time.

If the fibres enter the body, travel into the lungs by inhalation or into other vital organs when swallowed or ingested, they can become trapped. The fibres may embed themselves in the body tissue and can remain unaffected by natural digestion processes and body fluids for many years, sometimes permanently.

The damage and the continuous reaction of the body's defence mechanism against such foreign intruders can cause a number of serious, sometimes fatal diseases, such as:

ASBESTOSIS   fibrosis of the lung (this is a 'divisible disease', i.e., its development and course is affected by the cumulative impact of the frequency and severity of exposures to asbestos).

CANCERS   of the lung, larynx, ovary or other sensitive internal organs.

MESOTHELIOMA   a tumour of the chest or abdominal lining (this is an 'indivisible disease', i.e., it is believed that it can develop after a single exposure to asbestos).

Because of their different physical characteristics, the fibres of 'blue' (crocidolite) and 'brown' (amosite) asbestos are considered to be more harmful than those of 'white' asbestos (chrysotile). This is reflected in the regulatory controls that apply to the various types. Nevertheless, inhalation or ingestion of any form of asbestos presents serious health risks.

Asbestos-related diseases are virtually untreatable. There is no accepted safe level of exposure and even low levels can cause cancers to develop. Such cancers are often very painful and most people who contract mesothelioma die within one year of diagnosis.

The effects of exposure to asbestos are not immediate, and may not become apparent until 15–60 years or more after the first exposure. The risks increase progressively with continuous exposure. In addition, if a person is exposed to asbestos and is also a smoker, the risk of lung cancer is increased by more than 10 times.

In the UK in 1999 over 3,000 people died from asbestos-related diseases. The incidence is increasing. If it continues to increase at the same rate, there are fears that deaths could rise to 10,000 per annum by 2011.

Persons at risk

The greatest health risk is incurred by those who have worked in the asbestos industry – producing, processing, installing or removing the material. In addition, there is concern for people indirectly associated with such occupational exposure, such as those who lived in close proximity to a factory or processing plant, and family members of asbestos workers.

In the UK, the importation and use of almost all forms of asbestos is banned. Thus activities in which workers regularly have everyday contact with asbestos are now either prohibited or strictly controlled.

The greatest current risk is therefore to those who disturb asbestos, often inadvertently, incidentally in the course of their work, when inspecting, repairing, altering, extending or demolishing buildings that incorporate asbestos-containing materials (ACMs). Groups at risk in this respect include carpenters, plumbers, electricians, engineers installing air-conditioning or telecommunication equipment, shop-fitters and surveyors – in fact, anyone involved in the construction and property industry.

Building occupants are similarly at risk from any contamination arising from the disturbance or deterioration of materials or components containing asbestos, particularly where the airborne fibres are likely to be recirculated or distributed widely by ventilation plant and the like.

Assessing the risk

Newspapers are fond of using headlines such as 'one fibre kills'. However, as with the reaction to many hazardous products, some individuals are more susceptible to asbestos than others. It is widely accepted that the risk to health depends upon the type and number of asbestos fibres inhaled or ingested.

The factors to be considered are laid out in the table below.

Risk assessment table

Type	Blue and brown asbestos are more dangerous than white, because of their diameter, shape and other qualities. Asbestos cement is much less hazardous than asbestos insulating board
Asbestos content	Proportion of asbestos material as a percentage of the overall material
Condition	Intact or damaged
Friability	Density – firm or loose
Surface treatment	Sealed or unsealed
Dilution	Ventilated or unventilated air space
Location	Internal or external
Accessibility	Concealed, protected or exposed
Likelihood of future damage	By vandalism, children, animals, etc.
Building use	Type and age of occupants
Life expectancy	Age/exposure to deteriorating elements
Extent	Amount of material

Attitudes towards asbestos

All asbestos, in **any** circumstance, does **not necessarily** constitute a risk to health.

The policy of the Health and Safety Executive (HSE) on ACMs is that:

> '…asbestos materials which are in good condition and not releasing dust should not be disturbed….materials which are damaged, deteriorating, releasing dust or which are likely to do so should be sealed, enclosed or removed as appropriate following official guidance…

> …materials which are left in place should be managed and their condition periodically reassessed…

> …establish an order of priority in which remedial works should be undertaken…

> …substitute materials should be used where possible, provided they perform adequately …

> …the risk to the health of the public from asbestos materials which are in sound condition and which are undisturbed is very low indeed.'

Despite stringent controls, including the use of respirators, the health risks arising from the removal of asbestos are greater than from leaving it alone, provided it is not releasing or liable to release fibres.

Wherever asbestos remains, however, there is a risk of future disturbance. Moreover, while asbestos is extremely durable, the life of the material or component containing it is finite, and at some stage in the future the asbestos will inevitably need to be removed or replaced with a substitute.

Ingestion of asbestos

Although it cannot be proved that there is no risk, it is generally accepted that the health risks arising from the ingestion of asbestos fibres are very low, compared with their inhalation.

In buildings, the main source of fibres that could be ingested are the linings of water tanks and pipes, which may deteriorate due to the acidic content of the water.

Where cold water cisterns and pipes contain asbestos, this is in the form of asbestos cement. This form contains a relatively low proportion of asbestos and is usually of the low-risk white type. The risks that such pre-formed plumbing components present, due to natural deterioration, is generally very low. The greatest hazard arises if they are broken up for removal or are drilled, for example, for the installation of a new ball valve in a cistern. Even then, only rudimentary hygiene and protective measures are required to prevent the escape of any airborne fibres.

Appendix 3

Extent of asbestos-containing materials (ACMs)

Extent of ACMs

Asbestos has many and various useful properties, which have encouraged its use in buildings and plant. These include its great tensile strength; non-combustibility; resistance to heat, fire, electricity and chemical attack; its ability to be incorporated and to bind with other materials; and its relative cheapness and availability.

Consequently, it has been extensively used in almost all types of building – residential, educational, recreational, commercial and industrial – throughout the UK, in a wide variety of situations and forms.

Its useful properties have been appreciated and utilized for centuries, although the first asbestos-related deaths were not recorded until 100 years ago. The dangers to health have been known since the beginning of the 20th century, but were not seriously accepted by the industry until the late 1960s and the early 1970s.

Blue and brown asbestos continued to be used until they were banned in 1985. Asbestos insulating board was banned in the mid-1980s. White asbestos was not banned until November 1999. It is thought that as recently as 1998, £19m of asbestos-based roofing products alone were sold.

It is estimated that there are around 6m tonnes of asbestos remaining in buildings throughout the UK, and that a large majority of all commercial and industrial buildings contain asbestos in one form or another. A survey by the Association of Metropolitan Authorities (AMA) in 1985 estimated that 4m council homes in England, 8,000 of the 10,000 schools and 77% of the social services buildings surveyed (in AMA areas) contained some asbestos.

At approximately the same time, it was thought that there was over 40m sq metres of asbestos roof or wall cladding that was over 30 years old and nearing the end of its useful life.

Forms and products

Asbestos is an incredibly versatile material and has been used in a variety of materials and components. Those commonly found in buildings include the following, which are listed in approximate order of ease of fibre release.

Product	Use	Approx. % of asbestos
Loose	Mattresses/quilts for fire stopping or sound insulation	100
Sprayed coating	Dry or wet applied, anti-condensation or acoustic insulation, structural fire protection	55–85
Thermal insulation Lagging, pre-formed sections.	Of pipes, boilers, pressure vessels, calorifiers	6–85
Tapes, ropes, paper, felting, blankets		100
Asbestos boards		
Millboard	Fire protection, heat and electrical insulation	37–97
Insulating board	Fire breaks, infill panels, partitions, ceilings, ceiling tiles, linings to roofs and walls, external canopies and porch linings	16–40
Paper, felt, cardboard, electrical and heat insulation	Reinforcement and linings of other products	100
Textiles		
Ropes and yarns	Jointing and packing, boiler, oven and flue sealing – plaited tubing in electric cable	100
Cloth	Fire blankets, mattresses, curtains, gloves	100
Gaskets and washers	Hot water boilers to industrial power and chemical plant	90
Strings	Seals to hot water radiators	100
Friction products		
Resin-based materials	Bakes and clutch plates in machinery and lifts	30–70
Drive and conveyor belts	Engines and conveyors	30–70
Cement products		
Profiled sheets	Roofs, wall cladding, permanent shuttering	10–15
Semi-compressed flat sheet and partition board	Bath panels, soffits, walls, ceiling linings, weather boarding, composite panels for fire protection or as base for decorative facings	10–25
Fully compressed flat sheet for slates, tiles, board	Worktops, imitation roof slates	10–15
Pre-formed moulded and extruded products	Troughs and conduits, cisterns and tanks, drain pipes, flues, rainwater goods, windowsills and reveals, fascias, soffits, ducts, copings, promenade tiles and early imitation slates	10–15

Product	Use	Approx. % of asbestos
Other encapsulated materials		
Textured coatings	Decorative coatings on walls and ceilings	3–5
Bitumen products	Early forms of roofing felt, gutter linings and flashings, DPC, mastic and adhesives for floor tiles and wall coverings	100 (in bitumen matrix)
Flooring	Thermoplastic floor tiles	25
	PVC vinyl floor tiles and unbacked sheet	7
	Asbestos paper-backing to PVC floor tiles	100
Reinforced plastic and resin composites	Toilet cisterns, seats, banisters	1–10
Other	Asbestos is such a versatile material that not all of its uses are known. It is sometimes found in obscure, unexpected places	

Trade names

The Asbestos Information Centre Ltd has gathered together a useful (but not comprehensive) list of the trade names of ACMs. This is available for viewing on their website, www.aic.org.uk.

Appendix 4

Regulations

General

In view of the potential risks to health, all work that may disturb materials or components containing asbestos, including in particular their removal or treatment, is strictly controlled.

The relevant regulations are made under the enabling Act, the *Health and Safety at Work etc. Act* 1974. The two key regulations are the *Asbestos (Licensing) Regulations* 1983 and the *Control of Asbestos at Work Regulations* 2002 (*CAWR*).

Asbestos (Licensing) Regulations 1983

See the Health and Safety Executive (HSE) booklet L11, *A Guide to the Asbestos (Licensing) Regulations 1983 as amended.*

Broadly, these Regulations require any person who manages or carries out work with asbestos insulation, asbestos coating or asbestos insulating board, or conducts the clearance of asbestos-contaminated land where there is likely to be different types of asbestos-containing materials (ACMs), to hold a licence.

Licences are issued by the HSE following assessment of the competence of the applicant. They may be issued subject to terms and conditions and can be withdrawn if there is a serious transgression.

The Regulations do not apply to 'any article of bitumen, plastic, resin or rubber that contains asbestos and the thermal and acoustic properties of which are incidental to its main purpose'. Nor do they apply to 'asbestos cement' – hence the need to be able to distinguish between this and insulating board.

In addition, there are a few exemptions for 'minor works'*, namely where the work:

   i) is carried out by someone in the premises which they occupy, provided they have given the enforcing authority appropriate notice; or

   ii) is of short duration, i.e., lasting less than one hour in any period of seven consecutive days for any one person and less than two hours in total for all workers. (This does not include any time spent in preparatory work that does not disturb the asbestos); or

   iii) is solely air monitoring or sampling.

*The term 'minor works' does not appear in the Regulations or official guidance, but is used in this publication as a convenient way of avoiding the need to repeat these three circumstances, which in the context of the generality of asbestos work are only of minor interest.

Control of Asbestos at Work Regulations 2002

See the following HSE publications for information on these Regulations:

- L27 (4th edition) – *Work with asbestos which does not normally require a licence.* (*Control of Asbestos at Work Regulations*);

- L28 (4th edition) – *Work with asbestos insulation, asbestos coating and asbestos insulating board* (*Control of Asbestos at Work Regulations* 2002, Approved Code of Practice);

- L127 – *The management of asbestos in non-domestic premises.* (Regulation 4 of the *Control of Asbestos at Work Regulations* 2002, Approved Code of Practice and guidance).

Notwithstanding the requirement for a licence, these Regulations apply to any work that may involve the exposure of persons to any form of asbestos, irrespective of the type, form or amount, and include activities undertaken in both sampling and laboratory analysis.

Broadly, and using the number of the appropriate Regulations, they require:

a) The 'dutyholder' to:

4   manage asbestos in non-domestic premises;

b) 'Every person' to:

4.2 co-operate with the dutyholder so far as is necessary to enable him or her to comply with his or her duties to manage asbestos in non-domestic premises;

c) The employer to:

5   identify the type of asbestos;

6   prior to the works, assess the likely level of risk, determine the nature and degree of exposure and set out steps to prevent or control this;

7   produce a suitable written plan of work;

8   notify the enforcing authority of non-licensable work (at least 14 days before the works);*

9   inform, instruct and train personnel;

10   prevent exposure of employees to asbestos so far as is reasonably practicable, and where not reasonably practicable, reduce to the lowest level reasonably practicable both the exposure (without relying on the use of respirators) and the number of employees exposed;

11   ensure that any control measures are properly used or applied;

---

*These specific requirements are only triggered if the likely level of exposure to respirable fibres exceeds stated limits. The Regulations establish these trigger points in the form of 'action levels' and 'control limits'.

12 maintain control measures and equipment (keeping records of the latter);

13 provide suitable personal protective clothing and ensure that it is properly used and maintained;

14 establish arrangements to deal with accidents, incidents and emergencies;

15 prevent the spread of asbestos from the workplace, or where not reasonably practicable, reduce to the lowest level reasonably practicable;

16 keep asbestos working areas and plant clean, and clean thoroughly on completion;

17 designate 'asbestos areas' and 'respirator zones' and monitor and record exposure where appropriate;

18 monitor exposure of employees to asbestos by air monitoring* (if this is not deemed necessary, record the reasoning for this);

19 ensure that air testing is only carried out by a person with ISO 17025 accreditation;

20 ensure that the analysis of samples of material, to determine whether they contain asbestos, is only carried out by a person with ISO 17025 accreditation;

21 maintain health records and medical surveillance of employees;*

22 provide suitable washing and changing facilities (in addition to general welfare provisions); and

23 ensure that all raw materials and asbestos waste are stored, received into, dispatched from and distributed within suitable sealed and clearly labelled containers.

Regulations 4 and 20 do not come into force until 21 May 2004 and 21 November 2004, respectively.

Action levels

If the action level is likely to be exceeded, the following actions apply:

● written notification to the enforcing authority (14 days prior to commencing work);

● designation of 'asbestos working areas', where access is limited to authorized personnel;

● regular medical surveillance of employees, with health records kept for at least 40 years from the date of the last entry.

The action levels refer to the following accumulative exposures, whereby the respirable asbestos fibres per ml of air are multiplied by the number of hours of exposure over a continuous 12-week period.

---

*These specific requirements are only triggered if the likely level of exposure to respirable fibres exceeds stated limits. The Regulations establish these trigger points in the form of 'action levels' and 'control limits'.

Type of asbestos	Fibre-hours per ml of air
White	72
Any other type or mixture	48

Control limits

If the control limit is likely to be exceeded, the following actions apply:

- designation of 'respirator zones', where suitable respirators must be worn at all times; and

- the issue, use and maintenance of suitable respirators and protective equipment.

The control limits are as follows.

Type of asbestos	Average/measured in fibres per ml of air over a continuous period	
	4 hours	*10 minutes*
White	0.3	0.9
Any other type or mixture	0.2	0.6

Regulation 4: Duty to manage asbestos in non-domestic premises

The 'dutyholder' responsible for the management of asbestos in non-domestic premises, as set out in Regulation 4(1) of the *CAWR* 2002, is every person who by virtue of a contract or tenancy has an obligation for the repair or maintenance of the premises, or, in the absence of such, control of those premises or access thereto or egress therefrom. This includes those persons with any extent of responsibility for the maintenance or control of the whole or part of the premises.

When there is more than one dutyholder, the relative contribution required from each party in order to comply with the statutory duty will be shared according to the nature and extent of the repair obligation owed by each.

This Regulation does not apply to 'domestic premises', namely, a private dwelling in which a person lives. However, legal precedents have established that common parts of flats (in housing developments, blocks of flats and some conversions) are not part of a private dwelling. The common parts are classified as 'non-domestic' and Regulation 4 therefore applies to them, but not to the individual flats or houses in which they are provided.

Typical examples of common parts are entrance foyers; corridors; lifts and their enclosures and lobbies; staircases; common toilets; boiler rooms; roof spaces; plant rooms; communal services, risers and ducts; and external outhouses, canopies, gardens and yards.

The Regulation does not apply to kitchens, bathrooms or other rooms within a private residence that are shared by more than one household, or to communal rooms within sheltered accommodation.

(See Appendix 8 for a chart showing whether residential premises are likely to be classified as domestic or non-domestic for the purposes of Regulation 4.)

The duties of the dutyholder are to:

Co-operate*	Co-operate with other dutyholders so far as is necessary to enable them to comply with their Regulation 4 duties
Find and assess condition of ACMs	Ensure that a suitable and sufficient assessment is made as to whether asbestos is or is liable to be present in the premises, and as to its condition, taking full account of building plans and other relevant information, including the age of the building, and inspecting those parts of the premises that are reasonably accessible. (The dutyholder must presume that materials contain asbestos unless there is strong evidence to the contrary)
Review	Review the assessment in the event of significant change to the premises, or if it is suspected that it is no longer valid. Record conclusions of each review
Record	Keep an up-to-date written record of the location, type (where known), form and condition of ACMs
Conduct risk assessment	Where asbestos is or is liable to be present, assess the risk of exposure from known and presumed ACMs
Prepare and implement a management plan**	Prepare and implement a written plan, identifying those parts of the premises concerned and specifying measures for managing the risk, including adequate measures for properly maintaining asbestos or, where necessary, for its safe removal
Provide information to others	Make certain that the plan includes adequate measures to ensure that information about the location and condition of any asbestos is provided to every person likely to disturb it and is made available to the emergency services
Review and monitor	Regularly review and monitor the plan to ensure that it is valid and that the measures specified are implemented and recorded

*See sub-section 4.8 'Everyone (duty to co-operate)' for further details and advice.
**See section 10 ' Management plan' for further details and advice.

Storage, distribution and disposal of asbestos waste

Employers must ensure that raw asbestos or asbestos waste is stored, received into, dispatched and distributed in suitable sealed containers, clearly marked, in compliance with Schedule 2 of the *CAWR*, or where applicable, other more general regulations, including:

- the *Carriage of Dangerous Goods (Classification Packaging and Labelling) and the Use of Transportable Pressure Receptacles Regulations* 1996; and

- the *Carriage of Dangerous Goods by Road Regulations* 1996.

'Asbestos waste', defined as being an insulation material containing more than 0.1% asbestos (weight for weight) is subject to the waste management provisions of the *Control of Poisonous Waste Act* 1972 and the *Special Waste Regulations* 1996, enforced by the Environment Agency in England and Wales and the Scottish Environment Protection Agency in Scotland.

Disposal arrangements should be discussed and agreed with the appropriate disposal authority, which will be able to provide details of suitable licensed tips. The collection, delivery and disposal of the waste must be recorded using a 'consignment note', which must be signed for at each stage of the process, to ensure safe removal from the site and delivery to the approved final destination and to prevent fly-tipping.

Construction (Design and Management) Regulations 1994

Applicable works

Asbestos is specifically mentioned in the *Construction (Design and Management) Regulations* 1994 *(CDM)*, wherein the presence, location and condition of hazardous materials (such as asbestos) is included as an example of relevant information to be provided by the client to the planning supervisor for inclusion in the pre-construction health and safety plan.

Asbestos removal is, however, not one of the numerous activities that fall within the definition of 'construction work'. Such activity is therefore only included if it forms part of a project which, by virtue of the scope of other works, is applicable – for example, the refurbishment, alteration, dismantling or demolition of buildings or other structures.

Information

Regulation 11 of the *CDM* requires the client to ensure that, prior to the commencement of work, the planning supervisor is provided with any relevant information about the state or condition of the premises. This is extended to include that which could be ascertained 'by making enquiries which it is reasonable for a person in his position to make'. The *CDM* Approved Code of Practice (ACOP) clarifies this by noting that 'this may include surveys and other investigations'. Furthermore, asbestos is specifically mentioned as an example of a hazardous material, and the need for an early asbestos survey is reinforced in the working example of good practice given in the guidance.

These not so subtle indications of how the Regulations may be interpreted mean that clients who do not have a suitable asbestos survey undertaken at the earliest opportunity will need to be able to justify their decision if challenged.

Health and safety files

Regulation 14 (d, e and f) of the *CDM* requires the planning supervisor to ensure that at the completion of construction work a health and safety file is prepared for each structure. This must be passed to the client, who under Regulation 12 must retain and make the information available to anyone who might need to refer to it for future *CDM*-applicable activities. The purpose of this is to highlight any significant health and safety issues arising that may not be obvious.

The presence of asbestos, its treatment and any records of where asbestos has been previously removed is important information that should be incorporated within the file and updated as necessary.

Competence

Regulation 9 requires that only competent and properly resourced contractors and designers are appointed.

Other general regulations

Like any work operation, work with asbestos is subject to a whole raft of workplace health and safety legislation. Legislation with relevance to asbestos includes the following.

The Health and Safety at Work etc. Act 1974

This imposes a general duty on employers to conduct their works so as not to expose their employees or others to health and safety risks, and also to provide to other people information about the employer's workplace that might affect their health and safety.

Similar general duties apply to the self-employed and to anyone who has control, to any extent, over a workplace, its entrance or egress.

The Management of Health and Safety at Work Regulations 1999

These require employers and the self-employed to carry out risk assessments and to make appropriate arrangements for safeguarding themselves, their employees and others. They also require employers sharing workplace premises to co-operate in order to comply with health and safety legislation.

The Control of Substances Hazardous to Health Regulations 1999

These do not apply directly to asbestos, but do apply to other substances used in conjunction with asbestos working, such as sealants, adhesives and wetting agents.

The Defective Premises Act 1972 (in England and Wales) and The Government (Scotland) Act 1982 (in Scotland)

Both of these require landlords to see that tenants and others are not injured or infected by disease caused by a defect in their premises.

Enforcement

The organization responsible for enforcing legislation in respect of working activities involving asbestos is determined by the *Health and Safety (Enforcing Authority) Regulations* 1989. The enforcing authority is either the local authority or the HSE. The distinction is made according to the activity occurring in the premises in which the work is being carried out, as well as the nature and duration of the working activity itself.

However, the Regulations have certain provisos. If the work is not entirely internal, or will take longer than 30 working or 500 person days, or involves the installation, maintenance or repair of electricity or gas systems or fittings, or needs to be physically segregated (as would be the case with major asbestos removal or treatment works necessitating the erection of sealed working enclosures), then the HSE is the enforcing authority. In addition, the HSE is the stated authority for all activities in factories and railways and for construction and demolition generally.

The HSE is also the enforcing authority for work to which the *CDM* Regulations apply (see discussion of *CDM* Regulations above) and polices asbestos licences.

Penalties

The maximum penalty for offences under asbestos legislation depends on the breach of the act or legislation under which the prosecution is made, and on whether the hearing is in the magistrates court or the crown court.

The magistrates court can impose a fine up to £20,000. The maximum penalty per offence in the crown court is an unlimited fine and/or imprisonment for up to two years, following failure to comply with a prohibition notice.

Recently, concern has been expressed at the relatively low level of fines. In the future, it is highly likely that harsher penalties will be applied.

In addition, employees may be liable for substantial damages claims under civil law, and managers may be permanently disqualified from being directors of any company.

Under *The Defective Premises Act* 1972 in England and Wales and the *Government (Scotland) Act* 1982 in Scotland, any premises in such a state as to be prejudicial to health are a statutory nuisance (section 79 of the *Environmental Protection Act* 1990) and the local authority can serve an abatement notice on the owner or occupier.

Appendix 5

Flow chart to determine Regulation 4 'dutyholder'

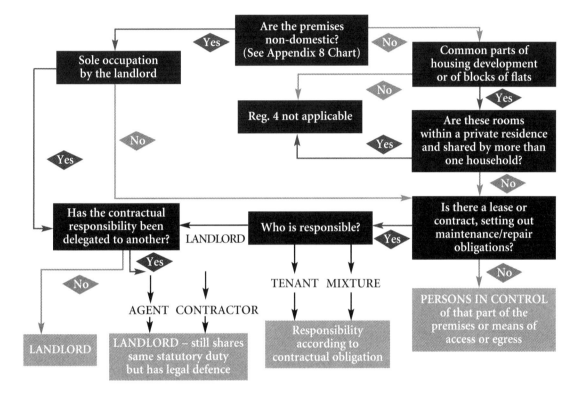

Appendix 6

Legal issues arising and cases

Is white asbestos safe?

There have been a number of recent articles in the national press that have suggested that the health risks arising from white asbestos (chrysotile) have been exaggerated, and that the material should not be regulated in the same way as brown (amosite) or blue (crocidolite) types.

Understandably, readers of such articles may be swayed by the arguments being promoted, including those relating to the financial cost of compliance.

The Health and Safety Executive (HSE) has responded to the issues raised by producing a '*Position Statement on the Risks From White Asbestos (Chrysotile)*'. In summary, this concludes that there is evidence that the health risk from white asbestos is less than for blue or brown asbestos, but that this type still carries a risk and remains a category 1 carcinogen.

HSE guidance (such as MDHS 100) allows risk assessments to take account of the type of asbestos found in a material. Both the 'action levels' and 'control limits' are less onerous for white than for other types. (See Appendix 4 for a description of these terms.)

Many building products contain a mixture of white and other forms of asbestos, giving them a greater overall risk than chrysotile on its own. For this reason, it is common practice for those engaged in asbestos work to assume that the asbestos found in a product is brown or blue and to take precautions accordingly. The *Control of Asbestos at Work Regulations* (*CAWR*) require this, where the type of asbestos is not known.

RICS members are not expected to have to decide between different and opposite expert scientific opinions. They must always follow the official guidance, i.e., that of the HSE, until directed otherwise.

The government has spent a lot of time and effort trying to educate property owners, occupiers, their advisors, building operatives and the general public about the hazards of disturbing asbestos. Attempting to differentiate between types without conclusive expert and official advice is a recipe for confusion and disaster. RICS members therefore must not get drawn into a debate on an issue on which they have no expertise, and nor should they anticipate or try to guess what may happen in the future.

Which employer is responsible?

Fairchild v Glenhaven Funeral Services

This legal case involved a claim for damages by an employee whose health was injured by exposure to asbestos in the course of his working life. The claimant had worked for two employers during the time his exposure to asbestos dust may have occurred. The employee had contracted mesothelioma, a fatal disease that is believed to be caused by a single exposure to asbestos (unlike asbestosis, in which the development and course of the disease is affected by the cumulative impact of the frequency and severity of exposures to asbestos).

The Court of Appeal initially ruled that the claim could not be made against either of the employers, unless it could be proven which employer was responsible for the injuring exposure. It was not fair to apportion blame on the grounds that at least one of them was guilty.

On 16 May 2002 a judicial committee of the House of Lords overturned this decision, on the basis that the breach of the employer's duty had materially contributed to a greater risk of the disease developing. They decided in this case that both employers were each 'jointly and severally liable'.

Each claim will be judged on the specific circumstances, but as a matter of public policy, courts are likely to follow the above principle, irrespective of the length of period of employment. As in the *Fairchild* case, where there has been more than one employer who has materially contributed to the risk, each will be deemed liable for the full damages and costs.

This is a ground-breaking change from the previous legal approach to causation and dramatically lowers the burden of proof required of claimants in future similar cases. The concern raised by many is that it may enable claimants to 'cherry-pick' the defendant who is most likely to settle or be able to afford the claim.

Registration of exposure to asbestos

There is usually a considerable delay (15 to 60 years) between actual exposure to asbestos fibres and the discernable symptoms of diseases appearing. Therefore, the incidence and/or extent of the injury, if any, cannot be proven at the time of exposure.

A successful criminal prosecution by the enforcing authorities enables a benchmark to be established that proves that a person has been exposed to an unacceptably high level of asbestos.

In British law, for a claim of negligence to be successful, damage or loss must be proven. However, depending on the circumstances, a claimant may be able to prove negligence, and have this officially recorded by the court, leaving damages to be assessed and successfully claimed at a later date, when the full extent of injury – if any – is known.

For the guilty party, the judgement remains hanging over them, and should be recorded as a possible financial matter in company accounts, and disclosed to insurers.

In certain states in the US, it is not necessary to prove actual harm. Judgements are made and damages (often punitive) are awarded based only on the degree of the claimant's exposure to asbestos.

Appendix 7

Competence of those involved in inspections for, and in the sampling, assessment and removal of asbestos

Asbestos inspectors

The British Occupational Hygiene Society (BOHS) controls various training modules for asbestos services, for which it issues proficiency certificates.

In conjunction with the Health and Safety Executive (HSE) and industry representatives, it has developed the following modules:

- S301* 'Asbestos and Other Fibres'

- P401 'Identification of Asbestos in Bulk Samples'

- P402 'Building Surveys and Bulk Sampling for Asbestos'

- P403 'Asbestos Fibre Counting (PCM)'

- P404 'Air Sampling and Clearance Testing for Asbestos'

*This was the original comprehensive module, the majority of the content of which has now been included within the more recent 'P40…' series of modules, according to the particular specialist areas of involvement.

Module S3O1 is aimed at persons who manage asbestos surveys or removal projects. Until recently, these courses were devised for asbestos consultants and contractors and were not widely known outside of the asbestos industry.

Following the introduction of the *Control of Asbestos at Work Regulations 2002* (*CAWR*), it is anticipated that thousands of skilled asbestos inspectors will be required to assist commercial tenants, building owners and those in control of premises in meeting the statutory duties that will be imposed on them. To this end, the United Kingdom Accreditation Service (UKAS) has devised an accreditation scheme for organizations carrying out asbestos surveys.

In addition, the BOHS and RICS (the latter in conjunction with the Asbestos Removal Contractors Association – ARCA) are both developing schemes for the personal certification of individual asbestos inspectors. All the schemes aim to establish a minimum standard of knowledge and experience, leading to certification in the form of a licence to practice, to be followed up by continuous monitoring to ensure that standards are maintained.

Neither accreditation nor certification are compulsory, but they will provide an industry-recognized benchmark and give clients guidance and reassurance in the selection of their inspectors.

The skills that an asbestos inspector will require are a detailed knowledge of asbestos and its health implications; an ability to conduct surveys in a wide variety of buildings of differing ages and construction; and the capacity to determine the presence or likelihood of the presence of asbestos and advise accordingly.

Specifically, the inspector will be required to be familiar with the different common types of asbestos and the health issues arising from each; the various forms in which asbestos was produced or used; the typical locations where it may be found within buildings, plant and machinery; and the means of assessing the risks arising.

The Approved Code of Practice (ACOP) (L127) accompanying the *CAWR* 2002:

- states that a different level of competence is needed for each of the three types of survey described in the Health and Safety Executive (HSE) guidance note MDHS 100 (see sub-section 8.3 'Types of survey');

- strongly advises that prior to employing anyone to carry out a survey, checks should be made to as to whether he or she has the relevant accreditation for the type of survey and practical experience of the type of building; and

- suggests that personnel working for an accredited organization or with appropriate personal certification are likely to be competent.

Analysts

Air-monitoring

Every employer who carries out analysis of the concentration of asbestos fibres in the air, or who requests another person to do so, must ensure that they are accredited to ISO 17025, or meet equivalent criteria (Regulation 19 of the *CAWR* 2002).

Analysis of bulk samples

From 21 November 2004, every employer who carries out analysis on a sample of material to determine whether it contains asbestos, or who requests a person to do so, must ensure that they are accredited to ISO 17025 or meet equivalent criteria (Regulation 20 of the *CAWR* 2002).

Accreditation

Currently, there is only one recognized accreditation body in the UK, the United Kingdom Accreditation Service (UKAS).

A full list of organizations accredited for undertaking various asbestos-related services, including surveying for, sampling, air-monitoring and analysis of bulk samples of asbestos-containing materials (ACMs), organized by region, is available from UKAS, 21–47 High Street, Feltham, Middlesex TW13 4UN (tel. 020 8917 8400; website www.ukas.com).

Asbestos removal contractors

The *Asbestos (Licensing) Regulations* 1983 effectively bar anyone from carrying out major works that may disturb asbestos unless they hold, or work in close association with someone who holds, a licence issued by the HSE. It must not be forgotten that even where a licence is not necessary, there will still be a requirement to comply with the *CAWR*.

Except for the simplest and for 'minor' works (see Appendix 4 'Regulations' for definition of this term), it is unlikely that unlicensed contractors will have the necessary resources, specialized equipment or the training and experience needed for work with this extremely hazardous material.

There are two trade organizations within the asbestos removal industry: the Asbestos Removal Contractors Association (ARCA), with the affiliated Asbestos Testing and Consultancy (ATAC) wing; and the Asbestos Control and Abatement Division (ACAD) of the Thermal Insulation Contractors Association (TICA).

A list of all licensed contractors, organized by region, is available from the HSE Asbestos Licensing Unit in Edinburgh.

Appendix 8

Chart showing whether residential premises are likely to be classified as domestic or non-domestic for the purposes of Regulation 4 of the CAWR 2002

The following information is taken either from the Approved Code of Practice (ACOP) to Regulation 4 of the *Control of Asbestos at Work Regulations* (CAWR) (indicated by an asterisk) or the response by the Health and Safety Executive (HSE) to questions put by RICS.

TYPE OF RESIDENCE	MODE OF OCCUPATION	ROOMS/PARTS	DOMESTIC PREMISES?	
			YES	NO
Private house	Owner/occupier	All	√	
Single dwelling including bed-sits	Let to single family	All	√	
	Occupied by more than one family	Private rooms (e.g. bedrooms and living rooms)	√	
	Multiple occupancy	Shared rooms* (e.g. kitchens, bathrooms and toilets)	√	
	Rooms let to lodgers	Common parts used for access and circulation (e.g. entrance lobby and staircase)		√

TYPE OF RESIDENCE	MODE OF OCCUPATION	ROOMS/PARTS	DOMESTIC PREMISES? YES	DOMESTIC PREMISES? NO
House converted into flats	Occupied by more than one family	Private rooms	✓	
	Occupied by more than one family	Common parts – for access, circulation and storage (e.g. entrance lobby and staircase or roof space)		✓
Garages/parking space	Integral/linked with residence	Private	✓	
	Not allocated to a specific person	Common parts		✓
Block of flats	Occupied by more than one family	Individual dwellings	✓	
	Occupied by more than one family	Common parts* (e.g. foyers, lifts, stairs, lobbies, boiler and plant rooms, roof spaces, communal yards, gardens, store rooms, external outbuildings and bicycle shelters)		✓
Flats over shop or office, with or without separate entrance	Occupied by owner of shop or office	Private rooms	✓	
	Leased separately	Private rooms	✓	
		Access/egress and circulation areas; private rooms		✓
Sheltered accommodation		Private rooms	✓	
		Communal rooms (dining rooms and lounge)	✓	
		Work areas (e.g. central kitchen, staff rooms and laundries), lifts, staircases and circulation areas, boiler room, stores and roof spaces		✓

TYPE OF RESIDENCE	MODE OF OCCUPATION	ROOMS/PARTS	DOMESTIC PREMISES?	
			YES	NO
Hotel or guest house including bed and breakfast accommodation when prime purpose, halls of residence, hostels owned privately or by the local authority, or care homes		Private rooms occupied by owner	✓	
		Guest accommodation and common parts		✓
Holiday homes*	Used substantially or exclusively as a business	All		✓
	Used exclusively by the owner and his or her family	All	✓	
Tied cottage/accommodation	Leased or rent-free	All	✓	
Farm house	Leased or rent-free	All	✓	

*If a holiday home is shared between domestic and commercial use, then the decision as to whether it is domestic or not will depend on the circumstances in each particular case, using the above principle as a basis for the assessment.

EFFECTIVE FROM 1 June 2003

Appendix 9
Bibliography

TITLE	TYPE	HSE ref	ISBN	CONTENT/AUDIENCE
Asbestos (Licensing) Regulations 1983	Reg.		0 11080 279 9	For persons regularly involved in works with asbestos-containing materials (ACMs) (asbestos removal contractors and those who manage such works)
Control of Asbestos at Work Regulations (CAWR) 2002	Reg.		0 71762 382 3	For anyone involved with ACMs in the workplace or in connection with their work
A Short Guide to Managing Asbestos in Premises		INDG 223 (rev 3)	0 7176 2 564 8	Free booklet – for potential dutyholders
A Comprehensive Guide to Managing Asbestos in Premises	Guide	HSG 227	0 71762 381 5	HSE guidance for those who have a duty to manage the risks from ACMs in premises. Designed for complex organizations or those responsible for older buildings likely to contain substantial quantities of asbestos and who need more guidance than that contained in *A Short Guide to Managing Asbestos in Premises*. The guide addresses the main factors that determine the risk presented by ACMs; notes the immediate steps to be taken to prevent exposure; and indicates how to develop a longer-term strategy and management plan. In addition, there are appendices on planning surveys, survey reports, material assessment and algorithms, worked examples of priority assessments, management options and the selection and management of asbestos contractors and analytical laboratories

TITLE	TYPE	HSE ref	ISBN	CONTENT/AUDIENCE
Work with asbestos which does not normally require a licence (4th edition)	ACOP	L27	0 71762 562 1	Contains *CAWR 2002*. Applies to work with asbestos that does not normally require an HSE licence, including building operations that disturb ACMs, asbestos sampling, laboratory analysis and work with asbestos during manufacturing. NB: this does not deal with Regulation 4 'Duty to manage asbestos in non-domestic premises' – see 'The management of asbestos in non-domestic premises' below
'The management of asbestos in non-domestic premises' (Regulation 4 of the *CAWR 2002*)	ACOP	L127		Advice on how to comply with Regulation 4 of the *CAWR 2002*. Designed for dutyholders and for every person required to co-operate with them
Work with asbestos insulation, asbestos coating and asbestos insulating board (CAWR 2002)	ACOP	L28	0 71762 563 X	Applies to works for which an asbestos licence is needed and for similar works by employers using their own employees on their own premises
Working with asbestos cement	Guide	HSG 189/2	0 71761 667 3	Notes precautions to be taken when working with asbestos cement products
Selection of suitable respiratory protective equipment for work with asbestos	Guide	INDG 288	0 71762 456 0	Provides advice on the selection of appropriate personal protection equipment (PPE)
MDHS 100: Surveying, sampling and assessment of asbestos-containing materials	Guide	MDHS 100	0 71762 076 X	Sets out how to survey workplace premises for ACMs, how to recognize and sample suspected ACMs, and how to record the results properly. For those carrying out or commissioning asbestos inspections
Introduction to asbestos essentials: Comprehensive guidance on working with asbestos in the building maintenance and allied trades	Guide	HSG 213	0 71761 901 X	For anyone who is liable to control or carry out maintenance work with ACMs that does not require an HSE licence. Contains information on where you are most likely to find asbestos and on appropriate protection methods. Includes photographs of typical ACMs; a drawing of a building, indicating typical locations for the most common ACMs; decision flow charts; and advice on waste-handling and the wearing of correct PPE. (See also *Asbestos Essentials – Task Manual* below)

TITLE	TYPE	HSE ref	ISBN	CONTENT/AUDIENCE
Asbestos Essentials – Task Manual	Guide	HSG 210	0 71761 887 0	Supplements *Introduction to asbestos essentials.* Particularly useful for workers who occasionally need to carry out minor works to ACMs. Contains task sheets giving detailed practical advice on 25 specific work activities, as well as eight guidance sheets on sundry associated issues, such as use of equipment and personal protection measures